CAMBRIDGE LIBRARY COLLECTION

Books of enduring scholarly value

Life Sciences

Until the nineteenth century, the various subjects now known as the life sciences were regarded either as arcane studies which had little impact on ordinary daily life, or as a genteel hobby for the leisured classes. The increasing academic rigour and systematisation brought to the study of botany, zoology and other disciplines, and their adoption in university curricula, are reflected in the books reissued in this series.

The Natural History of Birds

Georges-Louis Leclerc, Comte de Buffon (1707–88), was a French mathematician who was considered one of the leading naturalists of the Enlightenment. An acquaintance of Voltaire and other intellectuals, he work as Keeper at the Jardin du Roi from 1739, and this inspired him to research and publish a vast encyclopaedia and survey of natural history, the ground-breaking *Histoire Naturelle*, which he published in forty-four volumes between 1749 and 1804. These volumes, first published between 1770 and 1783 and translated into English in 1793, contain Buffon's survey and descriptions of birds from the *Histoire Naturelle*. Based on recorded observations of birds both in France and in other countries, these volumes provide detailed descriptions of various bird species, their habitats and behaviours and were the first publications to present a comprehensive account of eighteenth-century ornithology. Volume 3 covers corvids, thrushes and some tropical birds.

Cambridge University Press has long been a pioneer in the reissuing of out-of-print titles from its own backlist, producing digital reprints of books that are still sought after by scholars and students but could not be reprinted economically using traditional technology. The Cambridge Library Collection extends this activity to a wider range of books which are still of importance to researchers and professionals, either for the source material they contain, or as landmarks in the history of their academic discipline.

Drawing from the world-renowned collections in the Cambridge University Library, and guided by the advice of experts in each subject area, Cambridge University Press is using state-of-the-art scanning machines in its own Printing House to capture the content of each book selected for inclusion. The files are processed to give a consistently clear, crisp image, and the books finished to the high quality standard for which the Press is recognised around the world. The latest print-on-demand technology ensures that the books will remain available indefinitely, and that orders for single or multiple copies can quickly be supplied.

The Cambridge Library Collection will bring back to life books of enduring scholarly value (including out-of-copyright works originally issued by other publishers) across a wide range of disciplines in the humanities and social sciences and in science and technology.

The Natural History of Birds

From the French of the Count de Buffon

VOLUME 3

COMTE DE BUFFON
WILLIAM SMELLIE

CAMBRIDGE UNIVERSITY PRESS

Cambridge, New York, Melbourne, Madrid, Cape Town, Singapore,
São Paolo, Delhi, Dubai, Tokyo, Mexico City

Published in the United States of America by Cambridge University Press, New York

www.cambridge.org
Information on this title: www.cambridge.org/9781108023009

This edition first published 1793
This digitally printed version 2010

ISBN 978-1-108-02300-9 Paperback

THE

NATURAL HISTORY

OF

B I R D S.

FROM THE FRENCH OF THE

COUNT DE BUFFON.

ILLUSTRATED WITH ENGRAVINGS;

AND A

PREFACE, NOTES, AND ADDITIONS,

BY THE TRANSLATOR.

IN NINE VOLUMES.

VOL. III.

LONDON:

PRINTED FOR A. STRAHAN, AND T. CADELL IN THE STRAND:
AND J. MURRAY, Nº 32, FLEET-STREET.

MDCCXCIII.

CONTENTS

OF THE

THIRD VOLUME.

CONTENTS.

CONTENTS.

CONTENTS.

3. The

CONTENTS.

The

CONTENTS.

23. The

CONTENTS.

The

CONTENTS.

The

CONTENTS.

THE

N.° 57

THE CORNISH COUGH.

THE

NATURAL HISTORY

OF

BIRDS.

The RED-LEGGED CROW.

Le Crave ou *Le Coracias**, Buff.
Corvus-Graculus, Linn. and Gmel.
Gracula Pyrrhocorax, Scop.
Coracias, Aldrov. and Briff.
Coracias, feu *Pyrrhocorax*, Ray.
Cornix roftro pedibufque rubris†, Klein.
The Cornifh-Chough, *Cornwall-Kae*, *or Killegrew*, Alb. and Will.

SOME authors have confounded this bird with the Alpine Crow; but the diftinction is clearly marked. Its bill is longer, more flender, more hooked, and of a red colour; its tail is alfo fhorter, its wings longer, and, as a natural confequence, its flight is more lofty; and laftly, its eyes are environed by a fmall red circle.

It is true that the Red-legged Crow refembles the Alpine Crow in the colour and in fome com-

* In Greek, Κορακιας; and in modern Greek, *Scurapola:* in Cambden's Latin, *Avis Incendiaria:* in Italian, *Spelviero*, *Taccola*, *Tatula*, *Pazon*, *Zorl*, *Cutta:* in French, *Chouette* and *Choucas Rouge:* in German, *Stein-tahen* (*ftone-daw*), *Stein-tulen*, *Stein-krae*.

† *i. e.* The Crow with bill and feet red.

mon inftincts. In both, the plumage is black with green reflections of blue and purple, which have an admirable effect on that dark ground. Both delight in the fummits of the loftieft mountains, and feldom defcend into the plain. The former, however, is much more diffufed than the latter.

The Red-legged Crow is of an elegant figure, lively, reftlefs, turbulent, but can be tamed to a certain degree. At firft it is fed with a fort of pafte made with milk, bread, and grain, &c. and afterwards it is reconciled to whatever is ferved for our tables.

Aldrovandus faw in Bologna in Italy, a bird of this fort, which had an odd trick of breaking panes of glafs from the outfide, as if to enter the houfe by the window: this inftinct is undoubtedly the fame with that of the crows, the magpies, and daws, which are attracted by every thing that glitters. It has even been known to fnatch from the chimney light pieces of wood, and thus fet the houfe on fire; fo that this dangerous bird adds the character of an incendiary to that of a domeftic plunderer. But I fhould imagine that this pernicious habit might be turned againft itfelf, and, like the lark, it might be decoyed into fnares, by means of mirrors.

Salerne faw at Paris two Red-legged Crows which lived peaceably with the houfe pigeons; but it is probable that he had not feen the Wild Crow of Gefner, nor the defcription which that author

author gives of it; ſince he ſays after Ray, that it agreed in every thing but in ſize with the *coracias;* whether he meant the bird to which this article is allotted, or the *pyrrhocorax* of Pliny. Theſe birds are widely different, and Geſner was careful not to confound them. He knew that the Wild Crow differs from the Red-legged Crow by its creſt, its carriage, the ſhape and length of its bill, the ſhortneſs of its tail, the excellence of its fleſh, at leaſt when young; that it was not ſo noiſy or ſo ſedentary, and that it changed its reſidence more regularly at certain times of the year*; not to mention other differences.

The Red-legged Crow has a ſhrill though a pretty loud cry, very like that of the Sea-pie. It chatters almoſt inceſſantly; and Olina remarks that it is bred not for its voice, but for its beautiful plumage †. Belon, however, and the authors of the Britiſh Zoology ſay, that it learns to ſpeak.

The female lays four or five white eggs, ſpotted with dirty yellow. She builds her neſt on the tops of old deſerted towers, or on fright-

* "They arrive in the beginning of the ſpring, at the ſame time as the ſtorks.—They retire the firſt of all that I know, about the beginning of July," &c. GESNER *de Avibus.*

† The *cutta* with a red bill, which on other parts is all black like the Crow, except that its feet are yellow, comes from the mountains. In Latin it is named *Coracias.* This bird does not talk, but is kept merely on account of it beauty." *Uccelleria.*

ful precipices; for, according to Edwards, these birds prefer the cliffs all along the west coast of England, to similar situations on the flat shores of the east and south. I shall add another fact of the same kind, which I owe to a very respectable observer*. It is, that though these birds be inhabitants of the Alps, of the mountains of Switzerland, and of those of Auvergne, &c. they are never found on the mountains of Bugey, nor in all the chain that stretches along the confines of the country of Gex as far as Geneva. Belon, who saw them on Mount Jura in Switzerland, again observed them in the islands of Crete, and always on the summits of rocks. But Hasselquist affirms that these birds arrive in Egypt, and spread through the country after the inundation of the Nile has subsided and the waters are about to return into their bed. If we admit this fact, which however seems to be repugnant to the general nature of these birds, we must suppose that they are drawn to Egypt by the abundance of food with which the lands are replete, after being left by the waters to the powerful influence of a tropical sun: and in fact, they feed on insects, and on seeds which have been lately committed to the soil, and swell with milky juice, the effect of incipient vegetation. It follows then, that these birds do not confine their residence exclusively to rocks and the summits of mountains, since at

* Hebert, treasurer extraordinary of war at Dijon.

certain ſeaſons they regularly appear in Lower Egypt. Nor do they ſeem to be equally attached to every bleak eminence; but to be directed in their choice by certain peculiar circumſtances, which have hitherto eſcaped obſervers.

It is probable that the *coracias* of Ariſtotle* is the ſame with the Red-legged Crow, and not the *pyrrhocorax* of Pliny†, which ſeems to differ in ſize and in the colour of its bill, which is yellow. But the bird of which we here treat, has a red bill and red feet; and as it was ſeen by Belon on the Cretan mountains, it was more likely known to Ariſtotle, than the *pyrrhocorax*, which was ſuppoſed by the ancients to be confined to the Alps, and in fact was not ſeen by Belon in Greece.

I muſt admit, however, that Ariſtotle makes his *coracias* a ſpecies of daw (κολοιος), as we regard the *pyrrhocorax* of Pliny; which would ſeem to favour the identity, or at leaſt the proximity of theſe two ſpecies. But as in the ſame chapter I find a palmipede bird joined with the daws as of the ſame genus, the philoſopher evidently confounds birds that are of a very different nature; or rather, ſince the text traces a regular analogy, the confuſion muſt have ariſen from ſome miſtakes of the copyiſts. Beſides, the word *pyrrhocorax*, though entirely of Greek

* *Hiſt. Anim.* lib. ix. 24.

† Lib. x. 48.

 derivation,

derivation*, occurs not in any part of Ariſtotle's treatiſe; and Pliny, who was well acquainted with that work, could diſcover in it no account of the bird on which he beſtows that name; and in his deſcription of the *pyrrhocorax*, he does not copy what the Greek philoſopher had ſaid on the ſubject of the *coracias*.

The ſpecimen examined by the authors of the Britiſh Zoology weighed thirteen ounces, and its wings extended about two feet and a half: the tongue was almoſt as long as the bill, ſomewhat hooked; the nails black, ſtrong, and hooked.

Gerini mentions a bird of this kind whoſe bill and feet were black, and which he conſiders as a variety of the Red-legged Crow, affected only by ſome accidental differences of colour, ariſing from the diſtinction of the age or ſex†. [A]

* It ſignifies *fire-crow*.

† *Storia degli Uccelli*, tom. ii. p. 38.

[A] The ſpecific character of the Chough, *Corvus-Graculus*, Linn. is, that " it is blackiſh violet, its bill and feet red." It reſembles the jackdaw in habits and ſize, being ſixteen inches long: it is voracious, gregarious, and circles as it flies. Borlaſe ſays, that it is not as miſchievous as commonly repreſented, the tricks of the jackdaw being often imputed to it by miſtake.

The HERMIT CROW*.

Le Coracias Huppé ou *Le Sonneur* †, Buff.
Corvus-Eremita, Linn. and Gmel.
Coracia Criſtata, Briſſ.
Corvus Sylvaticus, Geſneri. Will.
Upupa Montana, Klein.
Geſner's Wood-Crow, Will.
Wood-Crow from Switzerland, Alb.

THIS bird is of the ſize of a hen; its plumage is black, with fine green reflections, which are variegated nearly as in the Red-legged Crow: like it, the bill and feet are red; but the bill is ſtill longer and more ſlender, very proper for inſerting into the fiſſures of rocks and the cracks in the ground, into the holes of trees and walls, in ſearch of inſects and worms, which are its principal food. In its ſtomach are found portions of the mole-crickets. It eats alſo the *larvæ* of the May-bug, and is uſeful on account of the havock which it makes among theſe deſtructive inſects.

The feathers on the top of its head are longer than the reſt, and form a kind of creſt, which

* In Zurich it is called *Scheller, Waldt-rapp, Stein-rap*; in Bavaria and Stiria, *Clauſs-rapp*; in Italian, *Corvo Spileto*; in Poliſh, *Kruk-leſny, Nocny*.

† i.e. *The Creſt Chough, or the Sounder.* This name has been given by ſome perſons to this bird, becauſe its voice reſembles the tinkling of the bells faſtened to the necks of cattle.

 hangs

hangs backwards; but this only appears after they are full grown, and again disappears when they are aged. Hence the reason that in some places they are called *Bald Crows*, and in some descriptions they are represented as having a yellow head marked with red spots. These colours are probably the tints of the skin, which age leaves bare.

The crest, which has given occasion to the name of *Mountain-crested* *, is not the only distinction between this bird and the Red-legged Crow; its neck is longer and more slender, its head smaller, its tail shorter, &c. Besides, it is known only as a bird of passage, while the Red-legged Crow, as we have already seen, is migratory, but only in certain countries and in particular circumstances. Gesner has therefore divided them properly into two species; and I have distinguished them by different names.

The Hermit Crows fly very lofty, and generally go in flocks †. They seek their food often in the meadows and marshy places, and always nestle on the tops of old deserted towers, or in the clefts of frightful inaccessible rocks. Sensible, as it were, that their young are delicate meat,

* Klein.

† I am aware that Klein makes the Hermit Crow a solitary bird; but this is directly contrary to what Gesner asserts, the only original observer, whom Klein copies, without being conscious, when he transcribes from Albin.

and

and much valued by the luxurious, they are careful to breed them out of the reach of man. But there are ftill fome men hardy enough to rifk their lives for the moft fordid gain, and allow themfelves to be let down by ropes from giddy heights, to plunder the infant brood in their receffes, and reap the moft dangerous of harvefts.

The females lay generally two or three eggs every hatch; and thofe who wifh to get the brood, commonly leave a young bird in each neft, in order to invite them to return the following year. When the young are plundered, the parents cry, *ka*, *ka*, *kæ*, *kæ*, but are feldom heard at any other time. The young are eafily tamed, and the more fo if they be taken early and before they can fly.

They arrive in the country of Zurich towards the beginning of April, at the fame time with the ftorks. Their nefts are fought for about Whitfunday, and they depart, the earlieft of all the birds, in the middle of June. I know not why Barrere has made the Hermit Crow a fpecies of curlew.

The Hermit Crow inhabits the Alps, the lofty mountains of Italy, Stiria, Switzerland, Bavaria, and the high cliffs which border on the Danube, in the vicinity of Paffau and Kelkeym. Thefe birds choofe for their retreat certain natural breaft-works, or cells of a good afpect, among the

the rocks, and hence the name *Klauſs-rappen*, or *Monk-Raven*. [A]

[A] Specific character: " Greeniſh, head yellowiſh, back of the head ſomewhat creſted, the bill and feet red."

M

No. 58

THE RAVEN.

The RAVEN.

Le Corbeau, Buff.
Corvus-Corax, Linn and Gmel.
Corvus, Briff. Klein, and Will.*
The Corbey, Sibb. *Scotia Illuſtrata.*

THIS bird has always been famous; but its bad reputation has been owing, moſt probably, to its being confounded with other birds, and loaded with their ill qualities. It has ever been regarded as the loweſt of the rapacious tribe; the moſt cowardly and the moſt diſguſting. Filth and rotten carcaſes, it is ſaid, are its chief food; and when it gluts its appetite on live prey, its victims are the weak or uſeful animals, lambs, leverets †, &c. yet it ſometimes attacks the large animals

* In Greek, Κοραξ: in Latin, *Corvus:* in Spaniſh, *Cuervo:* in Italian, *Corvo:* in German, *Rabe, Rave, Kol-rave, (coal-raven):* in Swediſh, *Korp:* in Poliſh, *Kruk:* in Hebrew, *Oreb:* in Arabic, *Gerabib:* in Perſian, *Calak:* in old French, *Corbin.* The appellations beſtowed, in all languages, are evidently formed from the Raven's croak. The Scotch name *Corbey,* like many others of that dialect, was introduced from the French. The Engliſh word *Raven* is derived from the German *Rabe.*—M. Montbeillard, author of this article, objects to the indiſcriminating application of the name *Corvus,* to the crows, daws, choughs, &c. The *corvus* of the ancients was appropriated to the large ſpecies, the Raven differing from the reſt conſiderably in its habits and inſtincts. It is as large as a good cock, and would weigh three carrion crows and two rooks.

† Aldrovandus relates a ſtory of two Ravens concerning an attack upon a hare; they picked out its eyes, and devoured it.

animals with ſucceſs, ſupplying its want of ſtrength and agility by cunning; it plucks out the eyes of buffaloes *, and then, fixing on the back, it tears off the fleſh deliberately: and what renders the ferocity more deteſtable, it is not incited by the cravings of hunger, but by the appetite for carnage; for it can ſubſiſt on fruits, ſeeds of all kinds, and indeed may be conſidered as an omnivorous animal †.

This violence and indiſcriminating voracity of the Raven has procured it a various treatment: ſometimes it has been proſcribed as a pernicious, deſtructive animal; ſometimes it has been afforded the protection of law, as uſeful in extirpating noxious inſects. In poor, thinly inhabited ſtates, the Raven may prove a burthenſome and expenſive gueſt; but in

* Ælian *Natur. Anim.* lib. ii. 51. *Recueil des Voyages qui ont ſervi à l'Etabliſſement de la Compagnie des Indes*, tom. viii. p. 273. This is, perhaps, the ſource of the antipathy which is ſaid to ſubſiſt between the Ox and Raven. *See* Ariſtotle *Hiſt. Anim.* lib. ix. 1. I can hardly believe that a Raven attacks a buffaloe, as travellers relate. It may happen that theſe birds will ſometimes alight upon the backs of the buffaloes, as the hooded crow alights upon the backs of aſſes and ſheep, or the magpie upon the backs of hogs, to eat the inſects which lodge in the hair of theſe animals. It may happen, too, that the Ravens, by exceſſive ſtrokes with their bill, may tear the buffaloes' hides, or even, attracted by the gliſtening of the pupil, they may pick out the eyes; but I cannot perſuade myſelf that they deliberately determine to eat the buffaloes alive, and are able to accompliſh that undertaking.

† *See* Ariſtotle *Hiſt. Anim.* lib. viii. 3. and Willoughby, p. 82. I have ſeen them fed, in a great meaſure, with fleſh, either raw or cooked.

rich,

rich, populous countries, it will be serviceable by devouring the filth generated in them. For this reason it was formerly, according to Belon, forbidden in England * to hurt this bird; but in the narrow islands of Ferroe, Malta, &c. a premium was offered for its destruction †.

If to the features which we have now traced of the Raven, we join its gloomy plumage; its cry, still more gloomy, though very feeble; its ignoble port, in proportion to its bulk; its savage look; its body smelling perpetually of infection ‡; we shall not be surprised that in all ages it has been regarded as an object of aversion and horror. Its flesh was forbidden

* Belon wrote in 1550. "This bird is esteemed sacred by our people, and few persons will dare to kill it." *Fauna Suecica*, N° 69.—The Ravens enjoy the same protection at Surinam, according to Dr. Fermin. *Description de Surinam*, tom. ii. p. 148.

† *Acts of Copenhagen* for the years 1671 and 1672. With regard to the island of Malta, I have been assured that the birds are carrion crows; but at the same time, as I am told they inhabit the most desert rocks on the coast, I am disposed to believe that they are Ravens.

‡ The authors of the British Zoology alone assert, that the Raven has an agreeable smell, which is difficult to believe of a bird that feeds on carrion. We know also by experience, that Ravens just killed, give the fingers a smell as disagreeable as that of fish. This I have been assured by Hebert, a very respectable observer; and the fact is confirmed by the testimony of Hernandez, p. 331. It is indeed said of the *Carancro*, a sort of vulture in America, which has also been termed a Raven, that it exhales an odour of musk, though it lives upon filth. (Dupratz *Hist. de la Louisiane*, tom. ii. 3.) But most authors assert directly the contrary.

to

to the Jews; ſavages never eat it *; and, among ourſelves, the moſt ſtarved wretches diſcover an extreme diſlike to it, and remove the coriaceous ſkin before they make their diſguſting meal. In every country it is reckoned an ominous bird, which announces impending calamities. Grave hiſtorians † have deſcribed pitched battles between armies of crows and thoſe of other ravenous birds, and have regarded theſe combats as foreboding the bloody wars kindled among nations. And how many perſons, at preſent, are alarmed and dejected at the noiſe of its croaking! The whole of its knowledge of futurity is limited, however, like that of the other inhabitants of the air, to a greater ſenſibility to the changes in its element, and to the expreſſion of its feelings by certain cries and actions. In the ſouthern provinces of Sweden, Linnæus tells us, that the Ravens, in fine weather, ſoar to an immenſe height, and make a clangorous noiſe, that is heard at a great diſtance §. The authors of the Britiſh Zoology add, that in this caſe they fly generally in pairs. Other writers ‖, in leſs enlightened times, have given other remarks, mingled with fable and ſuperſtition.

* Voyage du Pere Theodat. p. 300.

† Æneas Sylvius, *Hiſt. Europ.* cap. 53.—Bembo, *Init.* lib. v.—Geſner *de Avibus*, p. 347.

§ *Fauna Suecica*, No. 69.

‖ Pliny, Belon, Geſner, Aldrovandus, &c.

In

In thoſe times, when augury formed a part of religion, the Ravens, though bad prophets, could not fail to be birds of vaſt importance. The fondneſs of prying into futurity, how diſmal ſoever may be the proſpect, is an ancient malady of the human race. All the various motions of the Raven were ſtudied with the moſt ſcrupulous attention, all the circumſtances of its flight, all the differences of its voice, of which, not to mention the minute diſcriminations too difficult to be appretiated *, no leſs than ſixty-four diſtinct inflexions were reckoned up. Each had its determined ſignification; the artful applied themſelves to the profeſſion, and credulity drew multitudes to their oracles †. Pliny himſelf, though ſuperior to the prejudices of the vulgar, was ſo far carried by the tide of popular opinion as to mention its moſt infauſtous cries ‡. Some even carried this folly to ſuch lengths as to eat the heart and entrails of theſe birds, from the hope of acquiring the ſpirit of prophecy §.

But the Raven has not only a great number of inflections of voice correſponding to its interior affections, it has alſo the talent of imitating the cry of other animals ‖, and even human diſcourſe;

* Aldrovandus.

† Pliny, *lib.* xxix. *cap.* 4.

‡ *Id. lib.* x. 12. "The worſt omen is when they cluck with a ſtrangled voice."

§ Porphyr. *De abſtinendo ab Animant.* lib. ii.

‖ Aldrovandus.

and to improve this natural quality, the ligament of the tongue has been cut. *Colas* is the word which it pronounces the moſt eaſily*; and Scaliger heard one which, when hungry, called diſtinctly on the cook by the name of *Conrad* †. Theſe words bear indeed ſome reſemblance to the ordinary cry of the Raven.

Theſe ſpeaking birds were highly prized at Rome, and a philoſopher has not diſdained to relate the hiſtory of one of them ‡. They not only learn to prattle or repeat words, but become quite familiar. They can be tamed though old §, and appear even ſuſceptible of a laſting and perſonal attachment ‖.

In conſequence of their pliancy of temper, they can be inſtructed, not indeed to diveſt them-

* Belon.

† *Exercit in Cardanum.* Scaliger adds, as a pleaſant anecdote, that this ſame Raven, having found a paper with written muſic, pricked with its bill as if it were reading and beating time. It ſeems more natural to ſuppoſe that the bird miſtook the notes for inſects, on which it ſometimes feeds.

‡ Being early accuſtomed to ſpeak, it flew every morning to the *roſtra*, and ſaluted Tiberius, then the two Cæſars, Germanicus, and Druſus, and afterwards the Roman people as they paſſed by," &c. *Pliny*, lib. x. 48.

§ Geſner, p. 338.

‖ Witneſs the tame Raven mentioned by Schwenckfeld, which having ſuffered itſelf to be carried too far by its wild companions, and not being able to regain its way, diſcovered afterwards on the high road the man who had been uſed to feed it, hovered ſome time croaking above his head, as if to pay him homage, then alighted upon his hand, and never more left him. *Aviarium Sileſiæ*, p. 232.

ſelves

ſelves of their voracity, but to moderate it and direct it to the ſervice of man. Pliny ſpeaks of Craterus, an Aſiatic, who was noted for his ſkill in breeding Ravens to hunt, and who could make himſelf be followed even by the wild Ravens *. Scaliger relates, that king Louis (probably Louis XII.) had one ſo trained, and uſed it in the chace of partridges †. Albertus ſaw one at Naples which caught partridges and pheaſants, and even other Ravens; but to hunt birds of its own ſpecies it required to be rouſed, and, as it were, forced by the preſence of the falconer ‡. Laſtly, It can ſometimes be taught, it would ſeem, to protect its maſter and aſſiſt him againſt his enemies by its manœuvres: at leaſt if we give credit to the ſtory which Aulus Gellius tells of the Crow of Valerius §.

* Lib. x. 48.

† In Cardanum, exercit. 232.

‡ Aldrovandus, p. 702. Alſo Dampier, vol. ii.

§ A Gaul of high ſtature having challenged the braveſt Roman to ſingle combat, a Tribune named Valerius ſtepped forth and proved victorious by the aſſiſtance of a Raven which perpetually haraſſed his antagoniſt, tearing his hands with its bill, and darting at his face and eyes. Valerius afterwards bore the name of the *Raven (Corvus)*. *Noct. Atticæ*, lib. ix. 11.

[This ſtory is alſo related by Livy, lib. vii. 26. The Gauls were ſo much intimidated by the fall of their champion, that the Romans gained a complete victory. The Raven is ſaid to have perched on the head of Valerius, and was regarded as a token of victory ſent down from heaven: ſo that it muſt have been a wild Raven. But the ſtory is evidently fabulous. T]

 The

The Raven has alſo great ſagacity at ſcenting out * carrion from a diſtance: Thucydides aſcribes to it the inſtinct of abſtaining from the carcaſſes of animals that have died of the plague †. It has been ſaid alſo, that a bird of this kind, wanting to drink out of a veſſel which was too narrow to admit it, had the ſhrewdneſs to drop into it ſmall ſtones, which by degrees raiſed the water to the top ‡. This thirſt, if the fact be true §, is a circumſtance which diſtinguiſhes the Raven from all the reſt of the birds of prey, eſpecially from thoſe which feed on live game, which are ſtimulated by hunger, and never deſire but to drink blood. Another difference is, that the Ravens are more ſocial than the other rapacious birds: but it is eaſy to account for this; ſince, as they eat every ſort of food, and have more reſources than the reſt of the carnivorous kind, they can ſubſiſt in greater numbers on the ſame extent, and have not therefore the ſame cauſes of ſeparation. We may here obſerve, that though tame Ravens feed on all ſorts of fleſh, and thoſe in the ſtate of liberty be generally ſuppoſed to commit great havock among the moles and field-

* "The Ravens are the only birds employed in the auſpices that ſeem conſcious of their own faculties; for when the gueſts of Media were ſlain, they all flew out of Peloponneſus and the region of Attica." PLINY, lib. x. 12. from ARISTOTLE, lib. ix. 31. The ſame quality is alſo noticed in the Fauna Suecica, No. 69.

† Lib. ii.

‡ Pliny, lib. x. 43.

§ It is alſo remarked by Geſner.

mice;

mice *; Hebert, who has noticed them attentively for a long courſe of years, never ſaw them tear or mangle dead carcaſſes, or even ſettle upon them: he is therefore of opinion, that they prefer inſects, and eſpecially earthworms, to every other ſort of food. He adds, that earth is found in their excrements.

The Ravens, the real mountain Ravens, are not birds of paſſage, and in this reſpect they differ, more or leſs, from the Crows with which they aſſociate. They ſeem particularly attached to the rock where they were bred, or rather where they have paired; it is their ordinary reſidence, which they never entirely abandon. If they deſcend into the plains, it is to procure their ſubſiſtence; and this more rarely happens in ſummer than in winter, becauſe they avoid the heat, which appears to be the only influence that difference of ſeaſons produces on them. They do not paſs

* It is ſaid, that in the Iſle of France, a certain ſpecies of Raven is religiouſly kept, with a view to deſtroy the rats and mice. *Voyage d'un Officiér du Roi,* 1772. *p.* 122. It is ſaid that the iſlands of Bermudas having been ravaged five years in ſucceſſion by a prodigious multitude of rats, which devoured the plants and trees, and croſſed, by ſwimming, from one iſland to another; theſe rats ſuddenly diſappeared, and no cauſe could be aſſigned, except that a great number of Ravens had, in the latter years, reſorted to the iſlands, which were never ſeen there either before or ſince. But we have no proof that Ravens prey much upon rats: the inhabitants of the Iſle of France may, like others, entertain a prejudice; and with reſpect to the rats of the Bermuda iſlands, they may have been ſtarved to death; they may have devoured one another, as often happens; or they may have been drowned in their paſſage between the iſlands.

the night in the woods, like the Carrion Crows, they choose, in their mountains, a retreat sheltered from the northern blast, under the natural alcoves secured by the recesses and projections of the rocks. Thither they retire during the night, to the number of fifteen or twenty. They sleep perched on the bushes that grow between the rocks, and build their nests in the crevices, or in the holes of walls, on the tops of old deserted towers, and sometimes on the high branches of large straggling trees *. Each male attaches itself to a female, with which it remains united for the course of many years †; for these birds, which we view with disgust, can yet inspire mutual and constant love, and, like the turtle, express the gradual swell of passion. The male, if we believe some authors, begins always with a sort of love-song ‡, then caresses and bills with his mate; and it has even been alleged, that they copulate by the bill §. The fact is, that

* Linnæus says, that in Sweden the Raven nestles chiefly upon the pines. *Fauna Suecica*, No. 69. And Frisch asserts, that in Germany they pitch mostly upon great oaks; that is, they prefer the loftiest trees, whether pines or oaks.

† "They are said to maintain their conjugal engagements sometimes forty years."—ALDROVANDUS.—— Athenæus goes still farther.

‡ Oppian, *de Aucupio*.

§ Aristotle ascribes this absurdity to Anaxagoras, and is even at pains to refute it seriously; for the female Ravens, he says, have a *vulva* and *ovaria*, and that if the male semen entered by the mouth, it would be digested, and produce nothing. *De Generatione*, lib. iii. 6.

 we

we ſee their courtſhips frequently in the day-time; but the conſummation is performed in the ſilence and obſcurity of the moſt ſecret receſs *; and hence, probably, the origin of the fable. Nor muſt we aſcribe this to any motives of decency; wild animals are conſcious of the danger of their ſituation, and are anxious to provide for their ſecurity. The *White-John*, we have already ſeen, conceals itſelf while it drinks, becauſe, its head being plunged up to the eyes in the water, it is in danger of being ſurprized. The Raven has the more need of caution, ſince he is languid in the act of coition, which probably laſts a conſiderable time; he therefore ſeeks a ſecret retreat, where, in undiſturbed ſecurity, he may indulge his paſſion †.

The female is diſtinguiſhed from the male, according to Barrere, by its plumage being of a lighter black, and her bill weaker; and my own obſervations ſeem to confirm this remark. She lays, about the month of March ‡, five or ſix eggs §, pale and bluiſh green, marked with a great number of ſpots and ſtreaks of a dirty co-

* Albertus ſays, that he once witneſſed the copulation of Ravens, and that it was performed as uſual with other birds. See Geſner, p. 337.

† "The genus of the Ravens is not libidinous, becauſe it is not very prolific." ARISTOTLE, *de Generatione*, lib. iii. 6.

‡ Willoughby ſays that the Ravens ſometimes lay earlier in England.

§ Ariſtotle, *Hiſt. Anim.* lib. ix. 31.

lour*. She ſits about twenty days †, during which time the male provides her with food, and the ſupply is large; for the peaſants ſometimes find in the Ravens neſts, or near them, conſiderable heaps of grain, nuts, and fruits. It has been ſuſpected, indeed, that this hoarding is intended not only for the females during incubation, but for the ſubſiſtence of both through the winter ‡. But whatever be their motives, certain it is, that the Ravens ſteal not only proviſions, but whatever tickles their fancy, particularly bits of metal and glittering ſubſtances §. There was one at Erford, which had the aſſiduity to carry, one by one, and conceal beneath a ſtone in a garden, a quantity of ſmall pieces, amounting to five or ſix florins ‖. Every country furniſhes ſtories of ſuch domeſtic thefts.

When the young are hatched, they are far from being of the colour of their parents; they are rather white than black, contrary to the ſwans, which are originally brown, though deſtined to wear a ſnowy plumage ¶. At firſt the mother ſeems to treat her offspring with indifference, nor does ſhe feed them till they begin to be feathered: it has been alleged, that ſhe alters her conduct the moment ſhe is convinced by

* Willoughby.

† Ariſtotle, *Hiſt. Anim.* lib. vi. 6.

‡ Aldrovandus.

§ Friſch.

‖ Geſner, *de Avibus*, p. 338.

¶ Aldrovandus.

by their plumage that they are not ſpurious *. But for my part, I can ſee nothing in this that has not place in other animals, and even in man, ſome days after birth; a certain time is neceſſary to reconcile them to a new element and a new exiſtence. Nor is the young Raven then totally deſtitute of food; for a part of the yolk is included in the *abdomen*, and flows inſenſibly into the inteſtines by a particular duct †. After a few days, the mother feeds the young with the proper aliments, which previouſly undergo a preparation in her crop, and are then diſgorged into their bills, nearly as in the pigeons ‡.

But the male not only provides for the family, but watches for its ſafety. If he perceive a kite, or other ſuch rapacious bird, approach the neſt, the danger animates his courage; he takes wing, gains above his foe, and daſhing downwards, he ſtrikes violently with his bill; both contend for the aſcendency, and ſometimes they mount entirely out of ſight, till, overcome with fatigue, one or both fall to the ground §.

Ariſtotle, and many others after him, pretend that, when the young are able to fly, the parents drive them out of the neſt; and if the tract where they are ſettled affords too ſcanty a ſubſiſtence, they entirely expel them from

* Aldrovandus.

† Willoughby.

‡ *Idem.*

§ Friſch.

their precincts *. If this fact were true, it would shew that they are really birds of prey; but it does not agree with the observations which Hebert has made on the Ravens which inhabit the mountains of Bugey; for they protract the education of their brood beyond the period when these are able to provide for themselves. As it seldom happens that opportunity and talents concur in making such observations, I shall relate them in his own words:

" The young Ravens are hatched very early " in the season, and against the month of May " are able to quit their nest. A family of them " was every year bred opposite to my windows " upon the rocks which terminate the prospect. " The young, to the number of four or five, " sat on the large detached fragments about the " middle of the precipice, where they were ea- " sily seen, and drew notice by their continual " wailing. Every time that the parents " brought them food, which happened fre- " quently during the course of the day, they " called with a cry, *crau, crau, crau*, very dif- " ferent from their other noise. Sometimes one " tried to fly, and, after a slight essay, it returned " to settle upon the rock. Almost always some " one was left behind, and its wailing then be- " came incessant. After the young had strength " sufficient to fly, that is, fifteen days at least

* Aristotle, *Hist. Anim.* lib. ix. 31.

" after

" after their leaving the neſt, the parents con-" ducted them every morning to the field, and " in the evening led them back. It was com-" monly five or ſix in the afternoon when the " family returned, and they ſpent the reſt of " the day in noiſy brawling. This practice " laſted the whole ſummer, which would give " reaſon to ſuppoſe that the Ravens have not " two hatches annually."

Geſner fed young Ravens with raw fleſh, ſmall fiſhes, and bread ſoaked in water. They are very fond of cherries, and ſwallow them greedily, with the ſtones and ſtalks; they digeſt, however, only the pulpy part, and in two hours afterwards vomit up the reſt. It is alſo ſaid that they diſgorge the bones of thoſe animals which they eat entire, like the keſtril, the nocturnal birds of prey, the fiſhing birds, &c.* Pliny ſays †, that the Raven is ſubject every ſummer to a periodical diſtemper, which laſts ſixty days, whoſe principal ſymptom is exceſſive thirſt: but I ſuſpect that this is nothing but moulting, which is more tedious in this bird than in many others of the rapacious tribe ‡.

No perſon, as far as I know, has determined the age at which the young Ravens have acquired their full growth, and are able to propagate. If in the birds, as in the quadrupeds, each period of life was proportional to the total

* Aldrorandus. † Lib. xxix. 3. ‡ Geſner.

ſpace of exiſtence, we might ſuppoſe that the Crows required many years to reach their adult ſtate; for though the venerable age aſcribed by Heſiod * muſt be conſiderably curtailed, it ſeems well aſcertained that this bird ſometimes lives a century or more. In many cities of France they have been known to attain to that diſtant period; and in all countries and all ages, they have been reckoned as birds extremely long-lived. But the progreſs to maturity muſt be ſlow in this ſpecies compared to the duration of their life; for towards the end of the firſt ſummer, when all the family conſort together, it is difficult to diſtinguiſh the old from the young, and very probably they are capable of breeding the ſecond year.

We have already remarked that the Crow is not black at firſt. In the decline of life alſo, its plumage loſes the deep colour; and in extreme age, changes into yellow †. But at no time is this bird of a pure black, without the intermixture of other ſhades: Nature knows no ab-

* "Heſiod aſſigns nine of our ages to the Crow, the quadruple to the ſtags, and this tripled to the Ravens." PLINY, lib. vii. 48. If we eſtimate a generation at thirty years, the age of the Crow would be 270 years; that of the ſtag, 1080 years; and that of the Raven, 3240 years. The only way to give a reaſonable ſenſe to the paſſage, is to underſtand the *ætas* of Pliny, and the *γενεα* of Heſiod, to mean a year: and, on this ſuppoſition, the life of the Crow would be reckoned at nine years; that of the ſtag, thirty-ſix; and that of the Raven, 108, as proved by obſervation."

† Ariſtotle, *de Coloribus*.

ſolute uniformity. The black, which predominates, is mingled with violet on the upper part of the body, with cinereous on the throat, and with green under the body and on the quills of the tail, and the largeſt feathers of the wings and the remoteſt of the back*. Only the feet, the nails, and the bill, are quite black; and this colour of the bill ſeems to penetrate to the tongue, as that of the feathers appears to tincture the fleſh. The tongue is cylindrical at its baſe, flattened and forked near the tip, and roughened with ſmall points on the edges. The organ of hearing is very complicated, and more ſo, perhaps, than in the other birds†. It muſt alſo be more ſenſible, if we credit Plutarch, who ſays, that he has ſeen Crows fall down ſtunned with the noiſy acclamations of a numerous multitude, agitated by violent emotions ‡.

The œſophagus dilates at its junction with the ventricle, and forms a kind of craw, which was not overlooked by Ariſtotle. The inner ſurface of the ventricle is furrowed with wrinkles; the gall-bladder is very large, and adheres to the inteſtines ‖. Redi found worms in the cavity of the *abdomen* §: the length of the gut is nearly twice that of the bird itſelf, meaſuring

* Briſſon.

† Acts of Copenhagen, *ann.* 1673.

‡ Life of T. Q. Flaminius.

‖ Willoughby.—Ariſtotle, *Hiſt. Anim.* lib. ii. 17.

§ *Collect. Acad. Etrang.* tom. iv. p. 521.

from

from the tip of the bill to the extremities of the nails; that is, a medium between the extent of the inteſtines of the true carnivorous birds and the true granivorous: in a word, it is exactly ſuited for an animal which lives partly on fleſh, and partly on fruits *.

The appetite of the Raven, which is thus reconciled to every ſort of aliment, proves often its deſtruction, from the eaſe with which bird-catchers can provide a bait. The powder of the *nux vomica*, which is mortal to ſo many quadrupeds, is alſo a poiſon to the Raven; it is benumbed, and drops ſoon after eating the doſe; but the moment of intoxication muſt be ſeized, for the torpor is often only tranſient, and the bird recovers ſtrength ſufficient to reach its native rock, there to languiſh or expire †. It is alſo caught by various ſorts of nets, ſnares, and gins, and even by the bird-call, like the little warblers; for it alſo entertains an antipathy to owls, and cannot ſee them without venting a cry ‡. It is ſaid to wage war with the kite, the vulture, and the ſea-pie ‖; but this

* A reſpectable obſerver aſſured me, that he ſaw a Raven drop a nut more than twenty times from the height of twenty-four or thirty yards, and each time picked it up; but it could not ſucceed in breaking it; all this being done in a ploughed field.

† Geſner, p. 339. *Journal Economique*, Dec. 1758.

‡ Traite de la Pipee.

‖ Ælian. *Natur. Anim.* lib. ii. 51.—Aldrovandus, tom. i. p. 70. *Collect. Acad. Etran.* tom. i.

is

is nothing but the natural averſion to all carnivorous birds, which are enemies or rivals of each other.

When the Ravens alight upon the ground, they walk, but do not hop. Like the birds of prey, they have long vigorous wings, extending nearly three feet and a half; theſe conſiſt of twenty quills, of which the two or three firſt* are ſhorter than the fourth, which is the longeſt of all; and the middle ones have a remarkable property, viz. that the ends of their ſhafts ſtretch beyond the vanes, and terminate in points. The tail contains twelve quills, which are about eight inches long, but ſomewhat unequal, the two middle ones being the longeſt, then thoſe next, ſo that the end of the tail appears ſomewhat rounded on its horizontal plane †. This I ſhall afterwards call the *tapered tail* ‡.

From the extent of its wings we may infer the elevation of its flight. In ſtorms and tempeſts the Raven, it is ſaid, has been ſeen gliding through the air, conveying fire at its bill ||. This is only the luminous ſtar formed at the point of its bill, in its paſſage through the elevated regions of the atmoſphere, then ſurcharged with

* Briſſon and Linnæus ſay two; Willoughby, three.

† Add to this, that the Ravens have on almoſt their whole body a double ſort of feathers, ſo cloſely adhering to the ſkin, that they cannot be plucked without the help of hot water.

‡ *Queue etagee*; i. e. like the ſteps of a ſtair-caſe.

|| Scala Naturalis apud Aldrovandum, tom. i. p. 704.

with electricity. From ſome appearance of this kind, probably, the Eagle has been termed the miniſter of thunder; for there are few fables but are founded upon truth.

Since the Raven has a lofty flight, and is capable of enduring every temperature, the wide world is opened for its reception *. In fact, it is ſcattered from the polar circle † to the Cape of Good Hope ‡ and the iſland of Madagaſcar ||; and its number is determined by the quantity of food which the various intermediate regions ſupply, and the convenience of the ſituations which they afford §. It ſometimes migrates from the coaſts of Barbary to the iſland of Teneriffe. It is found in Mexico, St. Domingo, and Canada ¶, and undoubtedly in the other parts of the New Continent, and of the adjacent iſlands. When it is once ſettled in a country, and has become accuſtomed to its ſituation, it ſeldom quits it to roam into another **. It grows even attached to the neſt which it has built, and uſes it for ſeveral years together.

Its plumage is not the ſame in all countries. Beſide the changes which age introduces, the

* Aldrov. *Ornith.* † Klein.

‡ Kolben. || Flaccourt.

§ Pliny ſays, from Theophraſtus, that Ravens were ſtrangers in Aſia, Lib. x. 29.

¶ Charlevoix.

** Friſch.—Ariſtotle, *Hiſt. Anim.* lib. ix. 23.

colour

colour is alſo ſubject to vary from the influence of climate. It is ſometimes entirely white in Norway and Iceland, where numbers are alſo quite black *. On the other hand, white Ravens are found in the heart of France and Germany †, in neſts where ſome are likewiſe black. The Mexican Raven, called *Cacalotl* by Fernandez ‡, is variegated with two colours. That of the Bay of Saldagne has a white collar ‖; that of Madagaſcar, named *Coach*, according to Flaccourt, is white under the belly. The ſame mixture of black and white occurs in ſome individuals of the European ſort, even in what Briſſon terms the *White Raven of the North* §, which ought rather to have been called the *Black and White Raven*, ſince the upper part of its body is black and the under white, its head white and black, and alſo its bill, its feet, its tail, and its wings: theſe have twenty-one quills, and the tail has twelve; and what is remarkable, the

* Horrebow.—Klein. John of Cay ſaw in 1548, at Lubec, two white Ravens bred for the chace. *Id.* p. 58.

† *Ephemerides d'Allemagne.* Dr. Wiſel adds, that in the year following, black Ravens were found in the ſame neſts, and that in another neſt, in the ſame wood, a black Raven and two white ones were found. Of the latter colour, they are ſometimes killed in Italy. *See* GERINI *Storia degli Uccelli,* tom. ii. p. 33.

‡ *Hiſt. Avium Novæ Hiſpaniæ,* cap. clxxiv. p. 48. This is the *Corvus Varius* of Briſſon, and the *Red Raven* of Latham. Gmelin alſo reckons it a variety.

‖ Downton's Voyage, 1610.

§ This is the *White Raven* of Latham, which Gmelin makes a ſecond variety.

quills,

quills, at an equal diſtance on either ſide, which are commonly alike, are in this ſubject marked with black and white, differently diſtributed. This circumſtance would induce me to ſuppoſe that this is only an accidental change produced on the natural colour, which is black, by the exceſſive rigour of the climate; and if this conjecture be well founded, it would follow, that this is improperly reckoned a permanent ſpecies, eſpecially as all other animals that inhabit the arctic regions are clothed with a thicker fur than thoſe of the ſame kind which live in milder climates.

Theſe variations in the plumage of a bird ſo generally and ſo deeply impreſſed with black as the Raven, is another proof that colour can afford no permanent or eſſential character.

There is another kind of Raven which forms a variety in point of ſize. Thoſe of Mount Jura, for inſtance, appeared to Hebert, who had an opportunity of comparing them, to be larger than thoſe which inhabit the mountains of Bugey; and Ariſtotle * informs us, that the Ravens and Hawks were ſmaller in Egypt than in Greece. [A]

* *Hiſt. Anim.* lib. viii. 38.

[A] The ſpecific character of the Raven, *Corvus-Corax*, LINN. is, "That it is black; its back of a black ſky-colour; its tail ſomewhat rounded." To the very ample detail given in the text, we can add but few circumſtances. The Raven weighs three pounds, and is twenty-ſix inches long. In the northern countries of of Finmark, Iceland, and Greenland, it frequents the huts of the natives, feeds upon the offals of ſeals, and alſo devours birds egg . It whirls dextrouſly in the air, and changes its prey from bill to feet, for relief. It replies to the echo of its croak. The male ſits by day, and the female by night. On the approach of ſtorms, it gathers under the ſhelter of crags. The Greenlanders eat its fleſh, clothe themſelves with its ſkin, make bruſhes of its wings, and ſplit the quills for fiſhing-lines. When a phyſician, among the American ſavages, viſits a patient, he invokes the Raven, as the ſign of returning health: the Eſquimaux, however, deteſt and dread the whole genus.

M

FOREIGN BIRDS, RELATED TO THE RAVEN.

The INDIAN RAVEN of BONTIUS.

Buceros-Hydrocorax, Linn. and Gmel.
Corvus Indicus Bontii, Ray and Will.
Corvus Torquatus, Klein.
The Indian Hornbill, Lath.

THIS bird is found in the Molucca iſlands, and chiefly at Banda. Our knowledge of it is drawn from an imperfect deſcription and a wretched figure; ſo that we can only conjecture the European ſpecies to which it belongs. Bontius, the firſt and I believe the only one who has ſeen it, reckons it a Raven, in which he is followed by Ray, Willoughby, and ſome others; but Briſſon conceives it to be a Calao. I would rather adhere to the former opinion; and my reaſons are briefly theſe:

This bird, according to Bontius, reſembles the Raven in the ſhape of its bill and in its p rt; though its neck is rather longer, and a ſlight protuberance appears in the figure riſing on the bill.

This is a certain proof that he knew no other bird to which he could ſo readily compare it, and yet he was acquainted with the Calao of India.

India. He tells us indeed that it feeds upon nutmegs; and Willoughby conſiders this feature as different from the character of the Common Ravens; but we have already ſeen that theſe eat wild nuts, and are not ſo much carnivorous as generally ſuppoſed.

On the other hand, neither the deſcription of Bontius nor his figure diſcovers the leaſt trace of the indenting of the bill, which Briſſon regards as one of the characters of the Calaos; and the little bump which appears on the bill bears no reſemblance to the protuberance which diſtinguiſhes the Calao. Laſtly, the Calao has neither the ſpeckled temples, nor the black tail quills which are mentioned in the deſcription of Bontius; and its bill is ſo ſingularly ſhaped, that an obſerver could not, I ſhould ſuppoſe, have ſeen it, and not remarked its form, much leſs have taken it for the bill of a Common Raven.

The fleſh of the Indian Raven of Bontius has a pleaſant aromatic flavour, derived from the nutmegs, which conſtitute its principal food; and it is extremely probable that if our Raven had the ſame ſort of aliments, it would loſe its rank ſmell *.

It would require to have ſeen the Raven of the Deſert *(graab el zahara)*, which Dr. Shaw

* Linnæan character of Bontius' Indian Raven: "Its front bony, plain, and bare before, its belly yellow." It is often tamed, and employed to catch rats and mice: it is thirty inches long.

 mentions,

mentions *, to be able to refer it with certainty to its analogous European ſpecies. All that the Doctor ſays is, that it is rather larger than our Raven, and that its bill and feet are red. This laſt character has determined Dr. Shaw to reckon it a large Chough; that bird, as we have already ſeen, is indeed known in Africa; but how can we conceive a Chough to be greater than a Raven? I mention this to draw the attention of ſome intelligent traveller.

I find in Kœmpfer two other birds mentioned by the name of Ravens, without a ſingle character to juſtify that appellation. The one is, according to him, of a middle ſize, but extremely audacious; it was brought from China to Japan as a preſent to the emperor. The other, which was alſo given to the emperor of Japan, was a bird from Corea, exceedingly rare, and called *Coreigaras*; that is, the Raven of Corea. Kœmpfer adds, that the Ravens which are common in Europe are not found in Japan, no more than the parrots and ſome other birds of India.

[*Note.* We ſhould here place the Armenian bird, which Tournefort calls the *King of the Ravens*, if it were really a Raven, or belonged to that family. But a glance of the miniature figure will convince us that it is more related to the peacocks and pheaſants, by its beautiful creſt, its

* Shaw gives it alſo the names *Crow of the deſert, Red-legged Crow*, or *Pyrrhocorax*.

its rich plumage, its ſhort wing, and the ſhape of its bill, though it is ſomewhat longer, and though other ſlight differences occur in the form of its tail and of its feet. It is properly termed on the figure *Avis Perſica pavoni congener* (Perſian bird akin to the peacock); I ſhould therefore have mentioned it among the foreign birds analogous to the peacocks and pheaſants, if I had been earlier acquainted with it.]

The CARRION CROW*.

Le Corbine, ou *Corneille Noire*, Buff.
Corvus Corone, Linn. and Gmel.
Cornix, Gesner, Ray, Will. Klein, Briss. &c.

THESE birds spend the summer in the extensive forests, from which they occasionally emerge to procure subsistence for themselves and their infant brood. Their chief food in the spring is partridges eggs, of which they are very fond, and are so dexterous as to pierce them and carry them on the point of the bill to their young. The consumption is prodigious; and though they are not the most sanguinary of the rapacious tribe, we may reckon them the most destructive. Fortunately, they are not numerous; we should hardly find two dozen of pairs in a forest of five or six miles compass in the environs of Paris.

During winter they live with the Rooks and Hooded Crows, and nearly in the same way. In this season, numerous flights of all sorts of

* In Greek, Κορωνη, which name was also applied to the prow of a ship, from the resemblance to the Crow's beak : in modern Greek, Κȣρȣνα, Κȣρανα, Κομβα : in Chaldaic, *Kurka* : in Italian, *Cornice*, *Cornacchia*, *Cornacchio*, *Gracchia* : in Spanish, *Corneia* : in German, *Kraë*, whence the English name.

Crows

No. 59

THE CROW

Crows aſſemble about our dwellings, keeping conſtantly on the ground, ſauntering among our flocks and ſhepherds, hovering near the tracks of our labourers, and ſometimes hopping upon the backs of hogs and ſheep, with ſuch familiarity, that they might be taken for tame domeſtic birds. At night they retire into the foreſts to lodge on the large trees, which they ſeem to chooſe as the general rendezvous, whither they reſort from every quarter, ſometimes from the diſtance of three miles all round, and whence they again ſally out in the morning in queſt of ſubſiſtence.

But this mode of life, which is common to the three ſpecies of Crows, is not equally ſuited to them all; for the Hooded Crows and the Carrion Crows become exceſſively fat, while the Rooks continue always lean. But this is not the only difference that ſubſiſts; towards the end of winter, which is the ſeaſon of their amours, the Rooks remove into other climates, while the Carrion Crows, which diſappear at the ſame time in the plains, make only a partial flitting, and retire into the next large foreſts, where they diſſolve the general ſociety to form new connexions more endearing and more intimate. They form into pairs, and ſeem to divide their territory into diſtricts of about a quarter of a league in diameter, each of which maintains its ſeparate family*. It is ſaid that

* This has perhaps given occaſion to ſay, that Ravens expel their young from their diſtrict as ſoon as theie are able to fly.

 this

this connexion ſubſiſts inviolate during the reſt of their life; and it is even alleged that if one of the couple happen to die the ſurvivor will never enter into another union.

The female is diſtinguiſhed by her plumage, which is of inferior luſtre. She lays five or ſix eggs, and ſits about three weeks, during which time the male ſupplies her with food.

I had an opportunity of examining a neſt of a Crow which was brought to me in the beginning of July. It was found in an oak eight feet high, in a wood planted on a little hill, where were other oaks larger. The neſt weighed two or three pounds; it was formed on the outſide with ſmall branches and thorns rudely interwoven, and plaiſtered with earth and horſe-dung; the inſide was ſofter, and lined carefully with fibrous roots. I found in it ſix young already hatched, all alive, though they had eaten nothing for twenty-four hours; their eyes were not open*, and no plumage was to be ſeen on them except the point of the wing quills; their fleſh was a mixture of yellow and black; the tip of the bill and their nails yellow; the edges of the mouth a dirty white, and the reſt of the bill and feet reddiſh.

When a buzzard or keſtril approaches the neſt, the parents unite to attack them, and dart with ſuch fury that they often kill them, ſplitting the ſkull with their bills. They alſo fight with the

* See Ariſtotle *de Generatione*, lib. iv. 6.

ſhrikes;

ſhrikes; but theſe, though ſmaller, are ſo bold as often to prove victorious, drive them from the neſt, and plunder the young.

The ancients aſſert, that the Crows as well as the Ravens are watchful of their young after the period of their flight *. This ſeems to be probable, and I ſhould ſuppoſe that they do not ſeparate from their parents the firſt year; for theſe birds readily aſſociate with ſtrangers, and is it not natural to ſuppoſe that the ſociety which is formed in the ſame family will continue to ſubſiſt till interrupted by the breeding ſeaſon?

Like the Raven, the Crow can be taught to prattle; it is alſo omnivorous: inſects, worms, birds eggs, fiſh, grain, fruits, every thing, in ſhort, is ſuited to it. It breaks nuts by dropping them from a height †; it viſits ſnares and gins, and ſhares the plunder. It even attacks ſmall game when exhauſted or wounded, which in ſome countries has made it be bred for falconry ‡; but in its turn it becomes the prey of a more powerful enemy, ſuch as the kite, the eagle, owl, &c. §

* Ariſtotle, *Hiſt. Anim.* lib. vi. 6.

† Pliny, lib. x. 12.

‡ The nobility among the Turks keep ſparrow hawks, ſacres, falcons, &c. for the chaſe; others of inferior quality keep Grey and Black Crows, which they paint with different colours, carry upon their right hand, and call back by the ſound *houb, houb*, frequently repeated, *Villamont*, p. 677; and the Voyage to Bender by *the Chevalier Belleville*, p. 232.

§ "I myſelf ſaw a kite in the middle of winter plucking a crow near the high-way." Klein, *Ordo Avium*, p. 177.

Its

Its weight is ten or twelve ounces; it has twelve tail feathers, all equal, and twenty in each wing, of which the firſt is the ſhorteſt and the fourth the longeſt; its wings ſpread three feet; the aperture of the noſtrils is round, covered with a ſort of briſtles projecting forward; it has ſome black ſpecks round the eyelids; the outer toe of each foot is united to that of the middle at the firſt joint; the tongue is forked and ſlender; the ventricle ſomewhat muſcular; the inteſtines rolled into a great number of circumvolutions; the *cæca* half an inch long; the gall bladder large, and communicating with the inteſtinal tube by a double duct *. Laſtly, the bottom of the feathers, that is, the part which is concealed, is of a deep aſh-colour.

As this bird is exceedingly cunning, has an acute ſcent, and flies commonly in large flocks, it is difficult to come near it, and hardly poſſible to decoy it into ſnares. Some, however, are caught by imitating the ſcreech of the owl, and placing lime twigs on the high branches; or it is drawn within gun-ſhot by means of an eagle owl, or ſuch other nocturnal bird, raiſed on perches in an open ſpot. They are deſtroyed by throwing to them garden beans, of which they are very fond, and in which ruſty needles have been concealed: but the moſt ſingular mode of catching them illuſtrates the nature of the bird, which I ſhall for that reaſon relate.

* Willoughby.

A Carrion Crow is faſtened alive on its back firmly to the ground, by means of a brace on each ſide at the origin of the wings. In this painful poſture the animal ſtruggles and ſcreams; the reſt of its ſpecies flock to its cries from all quarters, with the intention, as it were, to afford relief. But the priſoner, graſping at every thing within reach to extricate himſelf from his ſituation, ſeizes with his bill and claws, which are left at liberty, all that come near him, and thus delivers them a prey to the bird-catcher *. They are alſo caught with cones of paper baited with raw fleſh; as the Crow introduces his head to devour the bait, which is near the bottom, the paper, being beſmeared with birdlime, ſticks to the feathers of the neck, and he remains hooded, unable to get rid of this bandage, which covers his eyes entirely; he betakes to flight, riſes almoſt perpendicularly into the air, the better to avoid ſtriking againſt any thing, till quite exhauſted, he ſinks down always near the ſpot from which he mounted. In general, though the flight of the Carrion Crows be neither eaſy nor rapid, they mount to a very great height, where they ſupport themſelves long, and whirl much.

This ſpecies has, like the Raven, varieties of white †, and of white mixed with black ‡, but which have the ſame inſtincts.

* Geſner.

† Schwenckfeld and Salerne.—Briſſon adds, that they have alſo the bill, the feet, and the nails white.

‡ Friſch.

Friſch

Frifch fays that he once faw a flock of fwallows travelling with a troop of variegated Crows in the fame direction. He adds, that thefe pafs the fummer on the coafts of the ocean, fubfifting on what the waves caft afhore; that in autumn they migrate to the fouth, never in large bodies, but in fmall divifions at certain intervals from each other; in which circumftance they refemble the Black Common Crows, of which they feem to be only a permanent variety.

It is very probable that the Crows of the Maldivas, mentioned by Francis Pyrard, are of the fame kind; fince that traveller, who faw them very diftinctly, remarks no difference. They feem however to be more familiar and bolder than ours; for they entered houfes to pick up whatever fuited them, and often the prefence of a man did not difcompofe them. Another traveller fays, that thefe Indian Crows, when they can get into a chamber, delight in doing all the mifchievous tricks that are afcribed to monkeys; derange the furniture, and tear it with their bills, overturn lamps, ink-ftands, &c. *

Laftly, according to Dampier, there are in New Holland and New Guinea † many Carrion Crows which refemble ours. There are alfo fome in New Britain; but it would feem, that

* *Voyage d' Orient*, du Pere Philippe de la Trinite, p. 379.

† According to this navigator the New Guinea Crows differ from ours only by the colour of their feathers, of which all that appears is black, but the ground white.

though

though there are many in France, England, and part of Germany *; they are ſcarce in the north of Europe. Klein mentions that they are rare in Pruſſia. They muſt be very uncommon in Sweden, ſince not even the name occurs in the enumeration which Linnæus has given of the birds of that country. Father Tertre aſſures us alſo that they are not to be found at all in the Antilles; though, according to another traveller, they are very common in Louiſiana. [A]

* *Page du Pratz.* Their fleſh is more palatable, he ſays, than in France, becauſe they do not feed upon filth, being hindered by the *carancros*, a kind of American vultures.

[A] Specific character of the Carrion Crow, *Corvus Corone*, LINN. "All of a ſky black, its tail rounded, its tail-quills ſharp." Its length is eighteen inches, its breadth twenty-ſix. It is more frequent in England than in any other part of Europe.

M

The ROOK*.

Le Freux, ou *La Frayonne*, Buff.
Corvus Frugilegus, Linn. and Gmel.
Cornix Frugilega, Briſſ. and Klein.
Cornix Nigra Frugilega, Ray. Will. and Friſch.

THIS bird is of an intermediate ſize between the Raven and the Carrion Crow, and it has a deeper cry than them. What diſtinguiſhes it the moſt, is a naked white ſkin, ſcaly and ſometimes ſcabby, that encircles the baſe of the bill, inſtead of thoſe black projecting feathers, which in the other ſpecies of Crows extend as far as the aperture of the noſtrils. Its belly is not ſo thick or ſtrong, and ſeems, as it were, raſped. Theſe differences, apparently ſo ſuperficial, imply more radical diſtinctions.

The peculiarities of the Rook reſult from its mode of life. It feeds upon grain, roots, and worms; and as in ſearch of its proper ſubſiſtence, it ſcratches deep in the ground with its bill, which in time becomes rough, the feathers at the baſe are worn off by the continual rubbing.

* In Greek, Σπερμολογος, or ſeed-gatherer; which is alſo the meaning of the Latin name, *Frugilega*: in German, *Roeck*; whence the Engliſh name and the Swediſh *Roka*: in Dutch, *Koore-Kraey*: in Poliſh, *Gawron*.

bing *. However, the ſtraggling feathers are perceived there; a ſufficient proof that the bird is not naturally bald.

The appetite of the Rook is confined to grain, worms, and inſects; it never prowls in the kennel, nor eats any ſort of fleſh: it has alſo the muſcular ventricle and the broad inteſtines of the granivorous tribe.

Theſe birds fly in numerous flocks, which are ſometimes ſo immenſe as to darken the air. We may conceive what havock theſe hordes of reapers will commit on newly-ſown fields, or on crops nearly ripe. Accordingly, in ſome countries government has interfered †. The Britiſh Zoology vindicates them from the aſperſion, aſſerting that they do more good than harm, by deſtroying the caterpillars that gnaw the roots of the uſeful plants, and blaſt the honeſt

* Daubenton the younger, keeper and demonſtrator of the cabinet of natural hiſtory, made an obſervation lately in a jaunt to the country, which relates to the preſent ſubject. This naturaliſt, to whom ornithology already owes ſo much, ſaw at a diſtance, in a field entirely uncultivated, ſix Crows of which he could not diſtinguiſh the ſpecies, which ſeemed very buſy lifting and turning over the ſtones that were ſcattered here and there, to get the worms and inſects lodged under them. They went ſo briſkly to work, that they made the ſmaller ſtones ſpring two or three feet. If this ſingular exerciſe, which no perſon before has attributed to the Crows, be familiar to the Rooks, it will afford another reaſon for the wearing and dropping of the feathers that encircle the baſe of their bill. In that caſe the name *turnſtone*, now applied to a ſingle ſpecies of bird, might become generic.

† Aldrovandus.

labours of the hufbandman. It would require a calculation to decide the point.

But not only the Rooks fly in flocks, they alfo neftle in company, as it were, with thofe of their own fpecies; and their fociety is very clamorous, efpecially when they have young. Ten or twelve nefts are fometimes found on the fame tree, and a great number of trees thus furnifhed occur in the fame foreft, or rather in the fame diftrict *. They feek not retirement and folitude, but rather prefer fettling near our dwellings. Schwenckfeld obferves, that they commonly prefer the large trees planted round cemeteries; becaufe perhaps thefe are frequented fpots, or afford worms in greater plenty; for we cannot fuppofe that they are attracted by the fcent of the dead bodies, fince we have already faid that they will not touch flefh. Frifch afferts, that if, in the breeding feafon, a perfon goes under the tree on which they are thus fettled, he will inftantly be deluged with their excrements.

One circumftance will appear fingular, though very like to what happens every day among animals of a different fpecies. When a pair are employed in conftructing their neft, one muft be left to guard it, while the other is procuring the fuitable materials; without this precaution, it is alleged, the neft would in an inftant be completely pillaged by the other Rooks which

* Frifch.

are

are ſettled on the ſame tree, each carrying off a ſprig to its own dwelling.

Theſe birds begin to build their neſt in the month of March, at leaſt in England*. They lay four or five eggs, ſmaller than thoſe of the Raven, but marked with broader ſpots, eſpecially at the large end. It is ſaid that the male and female ſit by turns. When the young are hatched, and able to eat, they diſgorge their food, which they keep in reſerve in their craw, or rather in a ſort of bag formed by the dilatation of the œſophagus †.

I find in the Britiſh Zoology, that after their hatch is over, they leave the trees where they neſtled; and that they return not again till the month of Auguſt, and only begin to repair or rebuild their neſts in October ‡. This would ſhew that they continue almoſt the whole year in England; but in France, in Sileſia, and in many other countries, they are undoubtedly birds of paſſage, if we except a few; the only difference is, that in France they announce the winter, while in Sileſia they are the forerunners of the ſummer §.

The

* Britiſh Zoology. † Willoughby.

‡ It is ſaid that the herons take advantage of their abſence to lay and hatch in their neſts. ALDROVANDUS.

§ See Schwenckfeld, p. 243. At Baume-la-Roche, which is a village of Burgundy, a few leagues from Dijon, ſurrounded with mountains and craggy rocks, and where the air is ſenſibly colder than at Dijon, I ſaw repeatedly in ſummer a flock of Rooks

 which

The Rook is an inhabitant of Europe according to Linnæus; but it would appear that there are ſome exceptions, ſince Aldrovandus is of opinion that there are none in Italy.

It is ſaid that the young ones are good eating, and that even the old ones are tolerable food when fat, but this is very rare. Country people have leſs averſion to their fleſh, knowing that they ſubſiſt not on carcaſes, like the Ravens and the Carrion Crows. [A]

which had lodged and neſtled above a century, I am aſſured, in the holes of rocks facing the ſouth weſt, and where it would be very difficult to get at their neſts, and not without being let down by cords. Theſe Rooks were ſo familiar that they ventured to ſteal the reapers' luncheons: they diſappeared about the end of ſummer a couple of months only, after which they returned to their uſual haunt. For theſe two or three years paſt they have not been ſeen, and their place was immediately occupied by the hooded Crows.

[A] Specific character of the Rook, *Corvus Frugilegus*, Linn. "Black, its front aſh-coloured, its tail ſomewhat rounded." The Rooks are remarked to fly chiefly in the morning and the evening. The *erucæ* of the dor-beetle (*Scarabæus melolantha*, LINN.) are what they ſearch for in the ground. They advance no nearer the pole than the ſouth of Sweden.

No. 60

THE ROYSTON CROW.

The HOODED-CROW.

Le Corneille Mantelée, Buff.
Corvus Cornix, Linn. and Gmel.
Cornix Cinerea, Briff.
Cornix Cinerea Frugilega, Gefn. and Ald.
The Royfton Crow, Ray. and Will. *

THIS bird is eafily diftinguifhed from the Carrion-Crow and the Rook by the colours of its plumage. Its head, tail, and wings are of a fine black, with bluifh reflections; it is marked with a fort of fcapulary of a greyifh white, which extends both ways, from the fhoulders to the extremity of the body. On account of this appearance, it has been called by the Italians, *Monacchia*, or *Nun*, and *Mantled Crow* by the French †.

It affociates in numerous flocks, like the Rook, and perhaps is ftill more familiar with man, preferring, efpecially in winter, the vi-

* This fpecies feems to have been unknown to the ancient Greeks and Romans. The moderns have given it the Latin appellations, *Cornix-Cinerea*, *Varia*, *Hyberna*, *Sylveftris*; *Corvus Semi-cinereus*: in Italian, *Mulacchia* or *Monacchia*: in Swedifh, *Kraoka*: in Polifh, *Vrona*: in German, *Holzkrae*, *Schiltkrae*, *Nabelkrae*, *Bundtekrae*, *Punterkrae*, *Winterkrae*, *Afskrae*, *Grauekrae* (i. e. Wood-Crow, Shield-Crow, Navel-Crow, Pied-Crow, Punctured-Crow, Winter-Crow, Afh-Crow, Grey-Crow).

† The name *Hooded-Crow* is common in Scotland. SIBBALD.

 cinity

cinity of our farms and hamlets, and picking up its food in the kennels and dunghills, &c.

Like the Rook alſo, the Hooded-Crows change their abode twice a-year, and may perhaps be conſidered as birds of paſſage; for we annually perceive immenſe flocks arrive near the middle of autumn, and depart about the beginning of ſpring, ſhaping their courſe towards the north; but we are uncertain where they ſtop. Moſt authors aſſert, that they paſs the ſummer on the lofty mountains *, and build their neſts in the pines or firs; it muſt therefore be on mountains uninhabited and little known, as in thoſe of the Shetland iſles, where they are actually ſaid to breed †. In Sweden alſo ‡, they neſtle in the woods, eſpecially among the alders, and lay commonly four eggs; but they never ſettle in the mountains of Switzerland, of Italy §, &c.

Though, according to moſt naturaliſts, it lives on every ſort of food, worms, inſects, fiſh ‖,

* Aldrovandus, Schwenckfeld, and Belon.

† Britiſh Zoology. The authors of that work add, that it is the only ſpecies of Crow found in thoſe iſlands.

‡ Fauna Suecica. § Aldrovandus.

‖ Friſch ſays, that they are expert at picking fiſh-bones, and that when water is let out from pools, they quickly perceive the fiſh that are left in the pool, and loſe no time in catching them. It is eaſy, therefore, to perceive that the Hooded-Crows may often frequent the ſides of water; but this was no ſufficient reaſon for terming them aquatic or marine crows.

and

and even putrid flesh, and, above all, on the products of milk *; and though these facts would rank it among the omnivorous tribe, yet as seeds of various kinds, mixed with small stones, are found in its stomach †, we may infer, that they are the nearest allied to the granivorous species; and this is another trait in their character common to the Rook. In other respects, they resemble much the Carrion-Crow; they have nearly the same size, the same port, the same cry, and the same flight; the structure of their tail, wings, bill, and feet; the disposition of their internal parts, are exactly the same ‡; and if any difference can be perceived, they incline to the nature of the Rook. They often associate together, and nestle on the trees §; both lay four or five eggs, eat those of small birds, and sometimes devour the helpless animals themselves.

* Aldrovandus. † Gesner and Ray. ‡ Willoughby.

§ Frisch remarks, that they place their nest sometimes on the tops of trees, and sometimes on the lower branches; which would imply, that they sometimes breed in Germany. I have lately ascertained that they nestle sometimes in France, and particularly in Burgundy. A flight of these Crows has resided constantly, for two or three years past, at Baume-la-Roche, in certain holes of rocks, possessed above a century by Rooks. One year, these Rooks not having returned, a flock of fifteen or twenty Hooded-Crows immediately occupied their scite, have since had two hatches, and are at present (26th May 1773) engaged with a third. This is another analogous feature of the two species.

Analogies ſo ſtriking to the Carrion-Crow and the Rook, would lead us to ſuſpect that the Hooded-Crow is only the hybridous offspring of theſe two ſpecies. If it were only a variety of the Carrion-Crow, why does it fly in flocks, and ſhift its abode twice a-year *? or, if it were merely a variety of the Rook, whence thoſe numerous relations which it bears to the Carrion-Crow? But this double reſemblance will be eaſily explained, if we admit it to be a croſs-breed, participating of the qualities of both. This opinion would appear plauſible to philoſophers who are accuſtomed to trace phyſical analogies; but it derives additional probability from the conſideration that the Hooded-Crow is a new family, entirely unknown to the ancients.

Friſch ſays, that the Hooded-Crow has two cries; the one hollow and well-known, the other ſhrill and ſomewhat reſembling the crowing of a Cock. He adds, that it is ardently attached to its young, and that if the tree on which its neſt is built be cut down, it will fall with it, rather than abandon its offspring †.

* "The Raven and Crow are always ſeen, and never migrate or abſcond." ARISTOTLE *Hiſt. Anim.* lib. ix. 23.

† A ſimilar ſtory is told by White, in his "Natural Hiſtory and Antiquities of Selborne." A pair of Ravens had, for a courſe of years, been accuſtomed to breed in an ancient oak: the tree was cut down in the month of February, when the birds were ſitting, and the dam was cruſhed by the fall.

Linnæus ſeems to apply to this bird what is ſaid in the Britiſh Zoology with reſpect to the Rook, that it is uſeful in deſtroying deſtructive inſects. But do they not themſelves deſtroy more grain than the inſects which they extirpate? In many parts of Germany a price is ſet on their head *.

They are caught in the ſame ſnares as are the other Crows. They are found in all the countries of Europe, but at different times. Their fleſh has a ſtrong ſmell, and is little uſed, except by the lower ſort of people.

I know not for what reaſon Klein ranged the *Hoexototl*, or the Willow-Bird of Fernandez, among the Crows, unleſs on the aſſertion of Seba, who, deſcribing this bird as the ſame with that mentioned by Fernandez, makes it as large as an ordinary pigeon, while Fernandez, in the very place quoted by Seba, ſays that the *Hoexototl* is a ſmall bird of the ſize of a ſparrow, having the ſong of the goldfinch, and being good eating †. This is not much like a Crow, and ſuch miſtakes, which are ſo pregnant in Seba's work, muſt only throw confuſion into the nomenclature of natural hiſtory. [A]

* Friſch.

† The Crows muſt be widely ſcattered, ſince they appear in the beautiful ſeries of birds which Sonnerat has brought from India, the Molucca iſlands, and the land of the Papoux. The individual mentioned in the text came from the Philippines.

[A] Specific character of the Hooded-Crow, (*Corvus-Cornix*, Linn.): " Aſh-coloured, its head, throat, wings, and tail black.

 It

It weighs twenty-two ounces; its length twenty-two inches, its alar extent twenty-three. It is a bird of paſſage in Great Britain, appearing in the beginning of winter, and returning with the Woodcocks. It is found as far north as the Feroe iſlands and Lapmark, where it continues the whole year, but chiefly on the ſea-coaſt, ſubſiſting on ſhell-fiſh. Its toes are very broad and flat, which enables it to walk on marſhy grounds."

FOREIGN BIRDS,

WHICH ARE RELATED TO THE CROWS.

I.

The SENEGAL CROW.

Corvus Dauricus, Gmel. and Pallas.
The White-breasted Crow, Lath. and Kolben.

To judge of this from its ſhape and colours, which is all that we know of it, we ſhould ſuppoſe it moſt analogous to the Hooded-Crow, and differing only becauſe its white ſcapulary is not ſo much extended. Some diſtinctions are alſo perceived in the length of its wings, the ſhape of its bill, and the colour of its feet. It is a new ſpecies, and little known *.

* This bird is deſcribed in Pallas's Travels, and Gmelin thus ſtates its ſpecific character: "It is black, its nape whitiſh, its neck and breaſt white." It is of the ſize of the Jackdaw, being twelve inches long. It arrives in numerous flocks early in the ſpring, in the neighbourhood of Lake Baikal, from China and the ſouthern parts of the Mogul Empire.

II.

The JAMAICA CROW.

Corvus Jamaicensis, Gmel.
Cornix Jamaicensis, Briss.
The Chattering-Crow, or Cacao-Walk, Ray, Sloane, and Lath.

This foreign Crow ſeems to be modelled from ours, only its tail and feet are ſmaller; its plumage is black, like that of the Carrion-Crow. In its ſtomach are found red berries, grain, and catterpillars; which ſhews that its ordinary food is the ſame with that of our Rook and our Hooded-Crow. Its ventricle is muſcular, and lined in the inſide with a very ſtrong coat. This bird abounds in the ſouthern part of the iſland, and never leaves the mountains, in which reſpect it reſembles our Raven.

Klein characteriſes this ſpecies by the largeneſs of its noſtrils; but Dr. Sloane, whom he quotes, ſays only, that they are moderately large.

It is obvious that it belongs to the Crows; but it would be difficult to refer it to any one ſpecies, ſince it unites the qualities proper to each, and differs from them all by its continual chattering. [A]

[A] Specific character: "Totally black;" it is eighteen inches long.

No. 61

THE JACKDAW.

The JACKDAWS.

Les Choucas, Buff.
Corvus-Monedula, Linn. and Gmel,
Cornix Garrula, Klein.
Graculus, ſeu *Monedula*, Geſner.
Monedula, ſeu *Lupus*, Aldrov. Ray, and Briſſ.
The Kae, Sibbald *.

THESE birds are nearly related to the Crows; to inſtitute a compariſon between them would therefore throw light on the hiſtory of both. As there are three ſpecies of Crows, the black (the Carrion-Crow), the cinereous (the Hooded-Crow), and the bald (the Rook); ſo there are three correſponding ones among the Jackdaws: a black one (the Daw, properly ſo called); a cinereous (the Chough); and a bald. The only difference is, that the laſt is of America, and has little black in its plumage. In general the Jackdaws are ſmaller than the Crows; their cry, at leaſt that of the two European ſpecies, the

* In Greek, Λυκος, Κολιος, Βωμολοχος: in Latin, *Lupus*, *Graccus*, *Gracculus*, *Monedula*, (which Scaliger derives from *moneta*, a coin, on account of this bird's diſpoſition to pilfer): in Spaniſh, *Graio*, *Graia*: in Italian, *Ciagula*, *Tattula*, *Pola*: in German, *Tul* or *Duhl*, *Thale* or *Dahle*, *Thaleche* or *Dahlike*: in Saxon, *Aelcke*, *Kaeyke*, *Gacke*: in Swiſs, *Graake*: in Dutch, *Kaw*: in Flemiſh, *Gaey*, *Hannekin*: in Swediſh, *Kaja*: in Turkiſh, *Tſchauka*.

only

only kinds known to us, is ſhriller, and has influence in the forming of their names; ſuch as *Choucas*, *Graccus*, *Kaw*, *Kae*, &c. But it appears that they have more than one inflexion of voice; for I am aſſured that they ſometimes call out *tian*, *tian*, *tian*.

They live upon inſects, grain, fruits, and even fleſh, though very rarely; but they will not touch filth, nor do they haunt the coaſts to pick up the dead fiſh and other carcaſes that are caſt aſhore by the ſea *. In this circumſtance they reſemble more the Rook, and even the Hooded-Crow, than the Carrion-Crow; but they approach the latter by the habit of ſearching and hunting for partridge eggs, of which they deſtroy great numbers.

They fly in large flocks, like the Rooks; like theſe, too, they form a ſort of cantonments, which are even more numerous, conſiſting of a multitude of neſts crouded upon one another, in a large tree, in a belfry, or in the ruins of an old deſerted caſtle †. The male and female, when once paired, remain a long time ſteadily united. When the genial ſeaſon returns, which awakens the ſenſibility of the animal frame, they eagerly court each other's ſociety, and prattle inceſſantly; they toy and kiſs, till they are worked up to a

* Aldrovandus.

† Belon, Aldrovandus, and Willoughby. They prefer the holes of trees to the branches.

fury

fury which can no longer be ſatisfied with the calmer joys: nor do they omit theſe preliminaries when reduced to the ſtate of captivity*. After fecundation, the female lays five or ſix eggs, marked with a few brown ſpots on a greeniſh ground; and, after the young are hatched, ſhe watches, feeds, and rears them with an affection which the male is eager to ſhare. In this reſpect the Jackdaw reſembles the Crows, eſpecially the common ſort: but Charleton and Schwenckfeld aſſert that it has two hatches in the year; which has never been affirmed of any of the Crows, though it well correſponds to the order of nature, the ſmall ſpecies being always the moſt prolific.

The Jackdaws are birds of paſſage, though they are not ſo well entitled to that appellation as the Rooks and Hooded-Crows, ſince a number of them continue in the country through the ſummer. The towers of Vincennes are at every ſeaſon ſtocked with them, and ſo are all old buildings which afford the ſame convenience and ſhelter; but in France there are always fewer in ſummer than in winter. Thoſe which migrate, form themſelves into great bodies, like the Rooks and the Hooded-Crows; ſometimes they join the ſame army, and continually chatter as they fly: yet they keep not the ſame periods in France as in Germany; for they leave Ger-

* Ariſtotle, *de Generatione*, lib. iii. 6.

many

many in autumn with their young, and appear not again till the ſpring, after having wintered with us; and Friſch was right in maintaining that they do not hatch during their abſence, ſince neither the Jackdaws nor other birds breed in winter.

With reſpect to their internal ſtructure, I ſhall only obſerve that they have a muſcular ventricle, and near its ſuperior orifice a dilatation of the œſophagus, which ſerves in place of a crop, as in the Crows, but that the gall-bladder is more elongated.

They can be eaſily tamed and taught to ſpeak. They ſeem fond of the domeſtic ſtate; but they are faithleſs ſervants, concealing the food which they cannot conſume, and ſecreting bits of money and jewels.

To complete the hiſtory of the Jackdaws, we have only to compare together the two kinds which are natives of Europe, and afterwards ſubjoin, as uſual, the foreign ſpecies and varieties.

THE COMMON JACKDAW * is of the ſize of a pigeon; its iris is whitiſh, it has ſome white ſtreaks under its throat, ſome dots of the ſame colour round its noſtrils, and ſome of an aſh colour on the hind part of the head and neck; the reſt is entirely black, which is deeper, how-

* *Choucas*, Buff.

ever,

ever, on the upper parts, and gloſſed ſometimes with violet, ſometimes with green.

THE CHOUGH * differs from the preceding, in being rather ſmaller, and perhaps leſs common; its iris is bluiſh, as in the Rooks; the prevailing colour of its plumage is black, without any cinereous mixture, and ſmall white points are obſerved round its eyes. But in every other reſpect they are exactly alike, and there is no reaſon to doubt that they belong to the ſame ſpecies, and would breed together.

We need not be ſurpriſed that birds ſo nearly related to the Crows, ſhould preſent the ſame varieties. Aldrovandus ſaw in Italy a Jackdaw with a white collar; this is probably the ſame with what is found in many parts of Switzerland, and which for this reaſon the Engliſh call the *Helvetian Daw* †.

Schwenckfeld had occaſion to ſee a white Daw, with a yellowiſh bill. Theſe white Daws are more common in Norway and the cold countries; in the temperate climates even, as in Poland, a ſmall white Daw is ſometimes found in the neſt of the black Daws or Choughs ‡: in this caſe the colour of the plumage does not depend on the

* *Chouc*, BUFF. This is the *Monedula Nigra* of Briſſon, which Linnæus makes to be a variety of the Common Jackdaw, *Corvus Monedula*.

† The *Monedula-Torquata* of Briſſon, which Linnæus regards as a variety. The *Collared-Jackdaw* of Latham.

‡ Rzacynzki.

influ-

influence of climate, but arifes from a natural defect; in the fame way as white Ravens are bred in France, and white Negroes born in Africa.

Schwenckfeld fpeaks: 1ft, of a variegated Daw, which refembles the true Jackdaw, except that its wings are white, and its bill hooked. 2. Another Daw, which is very rare, and differs from the common kind in its being croffed *. But thefe are, perhaps, individual varieties, or monftrous productions. [A]

* I had this year, in my court-yard, four tufted hens, of a Flemifh breed, which had the bill croffed; the upper mandible was very hooked, and at leaft as much fo as in the crofs-bill itfelf; the lower was almoft ftraight. Thefe hens could not feed on the ground fo well as others; it was neceffary that grain fhould be laid of a confiderable thicknefs.

[A] Specific character of the Jackdaw, *Corvus Monedula*, LINN.: "It is dufky, the back of its head hoary, its wings and tail black." The Jackdaw weighs nine ounces; its length thirteen inches, and its breadth twenty-eight. It is found as far north as Sondmor, and fometimes in the Feroe iflands; it leaves Smoland and Eaft Gothland immediately after harveft, and returns in the fpring with the ftares.—Mr. White, in his Natural Hiftory of Selborne, relates a fingular fact: That, in a neighbouring warren, the Daws fometimes breed under ground in the rabbit-burrows.

No. 62

THE ALPINE CROW.

The ALPINE DAW*.

Le Choquard, ou *Choucas des Alpes*, Buff.
Corvus-Pyrrhocorax, Linn. and Gmel.
Pyrrhocorax, Gesner, and Aldrovandus.
The Alpine Crow, Lath.

PLINY calls this bird *Pyrrhocorax*, which name alone includes an abridged description. *Korax*, which signifies Crow, marks the blackness of its plumage, as well as the analogy of its species; and *pyrrhos*, which means rufous or orange, denotes the colour of its bill, which, in fact, varies between yellow and orange; and also that of its feet, which are still more variable, since they were red in the subject examined by Gesner, and black in the one described by Brisson. That author mentions also their being sometimes yellow, and others relate that they are yellow in winter, and red in summer. These yellow feet and bill, which last is smaller than that of the Jackdaw, have made it be taken for the Blackbird, and termed the Great Alpine Blackbird. But if we draw a comparison, we shall find that it approaches much nearer to the Jackdaws, by the size of its body, the length of its

* In Swiss, *Alpkachtel*, *Wildtul* (*Alp-kae*, *Wild-Daw*): in German, *Bergdol*, *Alprapp*, (*Mountain-Daw*, *Alp-Raven*).

 wings,

wings, and even the ſhape of its bill, though ſlenderer, and by its noſtrils being covered with feathers, which are thinner, indeed, than in the Jackdaws.

In the article of the Red-legged Crow, or Corniſh Chough, I have ſtated the difference between theſe two birds; which Belon, and ſome others who have not ſeen them, have confounded together.

Pliny believed the *Pyrrhocorax* to be of the Alps *; but Geſner, who has accurately pointed out the diſtinction between it and the Red-legged Crow, ſays, that in certain parts of the country of the Griſons, this bird does not appear in winter; and in other parts that it is ſeen nearly the whole year, but that its favourite reſidence, where it ſettles in numerous flocks, is the ſummit of lofty mountains. Theſe facts reſtrict ſomewhat the opinion of Pliny, but at the ſame time they confirm it.

The Alpine Daw is of a middle ſize, between the Jackdaw and the Carrion-Crow: its bill is ſmaller, and more arched, than either; its cry is ſhriller, and more plaintive than the Jackdaw, and by no means agreeable †.

* *Hiſt. Nat.* lib. x. 48.

† Schwenckfeld ſays, that the *Pyrrhocorax*, which he terms alſo the *Night-Crow*, is noiſy, eſpecially during the night, and ſeldom appears in the day; but I am not certain if Schwenckfeld means the ſame bird as I do, by that name.

It

It lives chiefly upon grain, and is very deſtructive among the crops; its fleſh is very indifferent eating. The inhabitants of the mountains draw meteorological predictions from its manner of flying; if it riſes aloft, they lay their account for cold; if it keep near the ſurface, they expect mild weather. [A]

[A] Specific character of the *Corvus-Pyrrhocorax*, LINN.: "It is blackiſh, its bill yellowiſh, its feet black." It is fifteen inches long.

M

FOREIGN BIRDS,

WHICH ARE RELATED TO THE JACKDAWS.

I.

The MUSTACHIO DAW.

Corvus Hottentottus, Linn. and Gmel.
Monedula Capitis Bonæ Spei, Briſſ.
The Hottentot Crow, Lath.

THIS bird is nearly the ſize of the Blackbird; its plumage is of a gloſſy black, like the Jackdaw's; and its tail is proportionably longer than in any of them; all the feathers which compoſe it are equal, and the wings, when cloſed, do not extend half its length; the fourth and fifth are the longeſt of all, reaching two inches and a half farther than the firſt.

There are two circumſtances to be remarked in the exterior of this bird: 1. Thoſe long and flexible black hairs which ariſe from the baſe of the upper mandible, and which are twice as long as the bill; beſides many other hairs, ſhorter and ſtiffer, and pointing forwards, and ſpreading over this ſame baſe, as far as the corners of the mouth. 2. Thoſe long and narrow feathers inſerted

ſerted in the upper part of the neck, which play on the back, and form a ſort of mane *.

* Specific character: "It is greeniſh black, its tail equal, with very long muſtachios." Its length is eleven inches and a quarter.

II.
The BALD-DAW.

Corvus Calvus, Gmel.
The Bald-Crow, Lath.

This ſingular Daw correſponds to the Rook; the anterior part of its head is bare as in the Rook, and its throat is only ſhaded with a few ſtraggling feathers. Its relation to the Daws in general is marked by the length of its wings, the ſhape of its feet, its port, its bulk, and its wide noſtrils, which are almoſt round. But it differs becauſe its noſtrils are not covered with feathers, and are placed in a deep cavity on either ſide of the bill, and alſo becauſe its bill is broader near the baſe, and ſcalloped at the edges. I can ſay nothing with reſpect to its hiſtory. It has not even received a name in any treatiſe of ornithology. It is a native of Cayenne †.

† Specific character: "It is duſky-ferruginous, its front and top bald." It is rather larger than the Jackdaw, being thirteen inches long.

III.

The NEW GUINEA DAW.

Corvus Novæ Guineæ, Gmel.
The New Guinea Crow.

The natural place this bird ought to occupy is between our Daws and what I call *Colnud*. It has the figure of the Daws, the grey plumage of one of them, at leaſt, on the upper part of the body; but it is not ſo large, and its bill is broader in the baſe, in which it reſembles the *Colnud*. It differs from the laſt by the length of its wings, which reach almoſt to the end of its tail, and from the *Colnud* and the Daws by the colours of the under-ſide of the body, which conſiſt of black and white ſtripes, that extend to the wings, and which bear ſome reſemblance to thoſe in the variegated Wood-pecker *.

* The ſpecific character of this bird includes a full deſcription, which would only be an unneceſſary repetition of the text, and is therefore omitted

IV.
The PAPUAN DAW.

Le Choucari de la Nouvelle Guineé, Buff.
Corvus Papuensis, Gmel.
The Papuan Crow, Lath.

The prevailing colour of this bird (for we know only its surface) is an ash grey, deeper in the upper side, and lighter in the under, and diluting, by degrees, almost to white under the belly and the parts adjacent. There are only two exceptions to this uniformity of plumage: 1. A black ring, which surrounds the base of the bill, and extends as far as the eyes. 2. The great quills of the wings, which are of a blackish brown.

In this bird the nostrils are, as in the preceding, entirely covered with white feathers; the bill is nearly similar, except that the ridge of the upper mandible is not round as in the Jackdaws, but angular as in the *Colnud*. It bears other relations also to the last, and resembles it in the proportions of its wings, which extend no farther than the middle of its tail; in the smallness of its feet, and in the shortness of its nails. In short, we cannot hesitate to place it, as well as the preceding, between the Jackdaws and the *Colnuds*. Its length, reckoning from the point of the bill to the extremity of the tail, is about thirteen inches.

We owe this new ſpecies, as well as the preceding, to Sonnerat *.

* Specific character: "It is cinereous, its belly white, its wing-quills duſky blackiſh."

V.

The CAYENNE COLNUD †.

Corvus Nudus, Gmel.
The Bare-necked Crow, Lath.

I rank this bird after the Daws, though it differs from them in ſome reſpects, becauſe it is certainly more analogous to theſe, than to any birds of our continent.

Like No. II. it has a very broad bill at the baſe, and reſembles it alſo in another reſpect, in being bald; but this is in a different way, the neck being almoſt naked and featherleſs. Its head is covered from the noſtrils incluſively, with a ſort of black velvet cap, conſiſting of ſmall ſtraight feathers, ſhort, interwoven, and very ſoft to the feel; theſe are more ſtraggling under the neck, and much more ſo under the ſides and in the hind part.

The Colnud is nearly of the ſize of our Jackdaws, and we may add that it wears their livery; for its plumage is entirely black, except

† i. e. *Naked-Neck.*

ſome

ſome of the coverts and the wing-quills, which are of a whitiſh grey.

To look at the feet of the one which I obſerved, one would ſuppoſe that the hind-toe was forcibly turned back; but this is its natural poſition, and it can be directed forward occaſionally, as in the martins. I have even remarked that it is connected with a membrane to the inner toe in each foot. It is a new ſpecies *.

* Specific character: "It is black, having a cape waved with ſoft feathers, its neck thinly covered with feathers."

VI.

The PHILIPPINE BALICASE †.

Corvus-Balicaſſius, Gmel.
Monedula Philippenſis, Briſſ.
The Philippine Crow, Lath.

I cannot prevail upon myſelf to give this bird the appellation of a Daw, as Briſſon has done; ſince, from his deſcription even, it appears to differ widely from them. Its wings extend only fifteen or ſixteen inches, and it is ſcarcely larger than a Blackbird: its bill is thicker and longer in proportion than in any of the European Jackdaws; its feet ſlenderer, and its tail forked; laſtly,

† In the Philippines it is called *Bali-Caſſio*.

laſtly, inſtead of the ſhrill gloomy cries of the Jackdaws, it has a ſoft pleaſant ſong. Such differences would lead us to expect many more, when the bird is better known.

Its bill and feet are black; its plumage of the ſame colour, with green reflections; its colour, at leaſt, is the ſame, therefore, with that of the Jackdaw *.

* Specific character: "It is of a greeniſh black, its tail forked."

No. 63

THE MAGPIE.

The MAGPIE*.

La Pie, Buff.
Corvus-Pica, Linn. and Gmel.
Pica Varia & Caudata, Geſner, and Aldrov.
Pica Ruſticorum, Klein.
The Magpie, *Pianet*, or *Piot*, Will. Alb. &c.

THIS bird reſembles the crow ſo much in external appearance, that Linnæus has claſſed them under the ſame genus; and Belon remarks, that if the tail of the Magpie was ſhortened, and the white removed from its plumage, it would be really a crow. In fact, the Magpie has the bill, the feet, the eyes, and the general ſhape of the crows and jackdaws; it has alſo many of their inſtincts and habits, for it is omnivorous, living on all ſorts of fruits, and devouring even carrion †, robbing the ſmall birds' neſts of the eggs and the young, and ſeizing ſometimes the parents, either by an

* In Greek, Κισσα or Κιτ]α, and on account of its variegated plumage, Ποικιλις: in Latin, *Pica*, *Ciſſa*, and according to ſome, *Avis Pluvia*, (Rain-Bird); in wretched modern Latin, *Ajacia*: in Spaniſh, *Pega*, *Picata*, *Pigazza*; and in Catalonia, *Graſſa*: in Italian, *Gazza*, *Ragazza*, *Aregazza*, *Gazzuola*, *Gazzara*, *Pica*, *Putta*: in German, *Aeloter*, *Atzel*, *Aegerſt*, *Agerluſter*: in Flemiſh, *Aexter*: in Poliſh, *Stroka*: in Swediſh, *Skata*, *Skuira*, *Skara*: in Daniſh, *Skade*: in Norwegian, *Skior*, *Tunfugl*.

† KLEIN.—I ſaw one that ate greedily the ſkin of an orange.

open attack, or by ſurpriſing them while enſnared. One has been known to fall upon a blackbird, another to ſnatch a crab, but ſtrangled by the cloſing of the claws, &c.*

Its fondneſs for live fleſh has ſuggeſted the breeding it for falconry, like the ravens †. It commonly ſpends the warm ſeaſon paired with its female, and engaged in hatching and breeding its young. In the winter it goes in flocks, and approaches the hamlets, where it has greater reſources, which the ſeverity of the ſeaſon renders the more neceſſary. It is eaſily reconciled to the ſight of man, ſoon grows familiar in the houſe, and at laſt becomes maſter. I knew one which paſſed a day and night among a crew of cats, which it was ſhrewd enough to command.

It prattles nearly like the carrion-crow, and learns to imitate the cries of animals, and even the human voice. One is mentioned which could exactly mimic the calf, the kid, the ſheep, and even the notes of the ſhepherd's pipe: another repeated completely, the flouriſh of trumpets ‡. Willoughby knew many which could

* ALDROVANDUS.—It occaſions much diſorder in the catching of birds by calls, and dares, ſo to ſay, the fowler in his hut.

† Friſch.

‡ Plutarch relates that a Magpie, which amuſed itſelf with imitating human diſcourſe, the cry of animals, and the ſound of inſtruments, having one day heard a flouriſh of trumpets, became mute all of a ſudden, which ſurpriſed perſons who had been ac-

8 cuſtomed

could pronounce whole phrases. *Margot* is the word commonly given them, because they can the most readily articulate it; and Pliny assures us, that this bird is very fond of that sort of imitation, is pleased with repeating the words it has learned, studies patiently and earnestly to recal those it has lost, is overjoyed with the discovery, and sometimes dies of vexation, if its recollection fails it, or if its tongue refuses to pronounce a hard word *.

The tongue of the Magpie is like that of the raven, for the most part black. It alights on the backs of hogs and sheep, like the jackdaw, and searches after the vermin which infest these animals; with this difference, however, that the hog receives its civilities with complaisance; but the sheep, no doubt more delicate and sensible, seems to dread it †. It also snaps, very dexterously, the flies and other winged insects which come in its way.

The Magpie can be caught by the same snares, and in the same manner with the carrion-crow, and it is addicted to the same bad habits of stealing and hoarding up provisions ‡; habits

customed to hear it chatter incessantly; but they were much more surprised sometime after, when, quite unexpectedly, it broke silence, not to repeat its usual lesson, but to copy the notes and modulations of the trumpets. PLUTARCH.

* *Hist. Nat.* lib. x. 42.

† Salerne.

‡ I have verified this fact, by scattering before a tame Magpie pieces of money and little bits of glass. It was discovered also, that they

habits almoſt ever inſeparable in the different ſpecies of animals. It is imagined alſo to forebode rain, when it chatters more than uſually * On the other hand, many circumſtances concur to ſeparate it from the crows.

It is much ſmaller than even the jackdaw, not weighing more than eight or nine ounces; its wings are ſhorter, and its tail longer in proportion, and hence its flight is neither ſo lofty, nor ſo well ſupported. It never undertakes diſtant journies, but only flies from tree to tree, or from ſteeple to ſteeple. When on the ground, it is in a continual flutter, hopping as much as walking, and briſkly wagging its tail. It ſhews, in general, more reſtleſſneſs and activity than the crows; it is more malicious, and is diſpoſed to a ſpecies of raillery †. The female accordingly diſplays greater art and contrivance in the conſtruction of her neſt; whether becauſe ſhe is more ardent for the male ‡, and therefore more attached to its young, or becauſe ſhe is aware that many birds of rapine are

they conceal their theft with ſuch great care, that it was ſometimes difficult to make a detection; under the bed, for inſtance, or between the quilts.

* Aldrovandus.

† "I once ſaw a Magpie flying towards a bird, which happened to be tied to ſome place; and as it was reaching to eat a bit of fleſh, the Magpie removed the morſel with its tail. I thence concluded that the Magpie delights in tantalizing other birds." Avicenna *apud Geſnerum*.

‡ The ancients had this idea; for, from the name of the Magpie Κισσα, they formed the verb κισσαν, *to deſire*.

are forward to plunder its eggs and its brood, and, besides, that some of them are prompted to retaliate. She places her nest on the tops of the loftiest trees, or, at least, on high bushes *, and, with the assistance of the male, strengthens it on the outside with flexible twigs and worked mud, and environs the whole with a basketing of small thorny branches closely entangled, leaving only in the thickest and most accessible side, a small hole for entering. But not contented with safety alone, she seeks convenience; she lines the bottom of the nest with a sort of round mattress †, on which the young repose soft and warm; and though this lining, which is the true nest, be only six inches in diameter, the whole mass, including the thorny embrasure, is at least two feet every way.

But all these precautions are not sufficient to remove her anxiety and apprehensions: she is per-

* It commonly builds in the skirts of woods or on vineyards.

† "The Blackbird and Magpie spread a bedding under their young." Aristotle, *Hist. Anim.* lib. ix. 13. I take this opportunity to observe, that many writers have thought that the Κισσα of Aristotle is the same with our jay, because the Κισσα is described to hoard up acorns, which are the principal food of the jay; but it is admitted that this food is common to both the jay and the Magpie; and there are two characters peculiar to the jay, which Aristotle could not fail to have observed, viz. the two blue marks on the wings, and the sort of crest which the bird makes by occasionally bristling the feathers on the head. Since Aristotle is silent on these particulars, I conjecture that his Magpie and ours are the same bird, and likewise the long-tailed variegated Magpie which was new at Rome, and rare even in the time of Pliny. *Lib.* x. 29.

perpetually on the watch; if ſhe perceive a crow to approach, ſhe flies immediately to meet him, haraſſes him, and purſues him to a diſtance *. If the enemy be more formidable, a falcon for inſtance, or an eagle, yet will not fear reſtrain; ſhe ruſhes on danger with a temerity which is not always crowned with ſucceſs. Her conduct muſt ſometimes, however, be more conſiderate, if we believe what is alleged, that when ſhe ſees a perſon ſpying her neſt with too envious an eye, ſhe tranſports the eggs to ſome other place, either between her claws, or in a way ſtill more incredible †. Sportſmen tell ſtories no leſs ſtrange about her ſkill in arithmetic, though her knowledge extends not beyond the number five ‡.

She lays ſeven or eight eggs at each hatch, and breeds only once a year, unleſs the neſt be deſtroyed or deranged, in which caſe ſhe conſtructs another, and both parents exert them-

* Friſch.

† "To the underſide of a twig they glue two eggs with their excrements, and, having taken the burthen on their neck, they tranſport it thus equally poiſed." PLINY, *lib.* x. 33.

‡ Sportſmen ſay, that if a Magpie ſees a man enter a hut built at the foot of the tree on which it breeds, it will not go into its neſt till it has perceived the man leave the hut; that if two enter, and only one comes out, it will not be deceived by the ſtratagem, but will ſtay till the ſecond alſo comes out; that it will likewiſe diſtinguiſh three out of four, and even four out of five, beyond which it cannot diſcriminate. It would thence follow, that the Magpie has a diſtinct idea of the ſeries of numbers, from one to five; and it muſt be confeſſed that the glance of a man's eye is not perhaps more accurate.

ſelves

ſelves with ſuch ardour as to complete it in leſs than a day. They have afterwards a ſecond hatch, conſiſting of four or five eggs; and if they be again diſturbed, they will rebuild the neſt, and make a third hatch, though ſtill ſmaller than the preceding *. The eggs of the Magpie are never ſo large, or of ſo deep a colour, as thoſe of the Raven; they are marked with brown ſpots, ſtrewed on a blue-green ground, moſt crowded about the thick end. John Liebault, quoted by Salerne, is the only one who affirms that the male and female ſit alternately.

The Magpies are hatched blind and ſhapeleſs, and it is ſome time before they aſſume their deſtined form. The mother not only rears them with an anxious care, but takes an intereſt in them after they are grown up. Their fleſh is indifferent eating, though it is not held in ſuch averſion as that of young crows.

With reſpect to the difference remarked in the plumage, I conceive it to be not abſolutely ſpecific; ſince, among the ravens, the crows, and the jackdaws, individuals are found variegated, like the Magpie, with black and white: we muſt admit, indeed, that in the former black

* Something of this kind has made the ſtratagem be imputed to the Magpie, of having conſtantly two neſts, with a view to diſappoint the birds of prey, by ſhifting from the one to the other. It was for the ſame reaſon that the tyrant Dionyſius had thirty bed-chambers.

is the ordinary colour, as the mixture of white and black is in the latter. But this is not uniform; and if we examine the bird closely, or view it in certain lights, we may perceive shades of green, purple, and violet, though not expected in a bird so little celebrated for beauty of plumage *. The male is distinguished from the female by the deeper blue gloss on the upper part of its body, and not by the blackness of its tongue, as some have alleged.

The Magpie is subject to moult, like the other birds; but it is observed, that the feathers drop successively and by degrees, except those on the head, which are detached all at once, so that at the annual return of the season it appears bald †. The young ones do not get their long tail before the second year; and, no doubt, this is the time when they become adult.

All that I can learn with respect to the duration of the life of the Magpie is, that Dr. Derham kept one twenty years, when it grew blind with age ‡.

This bird is very common in France, England, Germany, Sweden, and in every part of Europe, except Lapland ‖; it is also rare in mountainous countries, which shews that it

* British Zoology. † Pliny, *lib.* x. 29. ‡ Albin.

‖ Fauna Suecica, No. 76. M. Hebert assures me, that no Magpies are seen in the mountains of Bugey, or even at the height of Nantua.

cannot

cannot ſupport exceſſive cold. I ſhall cloſe this account with a ſhort deſcription, which will illuſtrate what cannot be repreſented at all, or which the figure expreſſes imperfectly.

It has twenty quills in each wing, the firſt of which is very ſhort, and the fourth and fifth the longeſt; twelve unequal quills in the tail, diminiſhing always in length, the farther they are from the two middle ones, which are the longeſt of all; the noſtrils round; the internal eye-lids marked with a yellow ſpot; the edges of the chops beſet with hairs; the tongue blackiſh, and forked; the inteſtines about twenty-two inches long; the *cæcums* half an inch; the *æſophagus* dilated and covered with glands at its junction with the ventricle, which is but little muſcular; the kidney oblong; and the gall bladder of the ordinary ſize *.

I have already ſaid that there are white Magpies as well as white ravens; and though the principal cauſe of this change of the plumage is the influence of northern climates, as may be ſuppoſed of the white Magpie of Wormius †, which was brought from Norway, and even

* Willughby.

† *Muſæum Wormianum*, p. 293. "It was ſent me from Norway, where two young ones of this kind were found in the neſt. It is evidently, from its form, a Magpie, only of a white colour, and ſmaller, not being yet full grown. Its head appears ſmooth."

 of

of ſome of thoſe mentioned by Rzacynſki *: I muſt confeſs, however, that they are ſometimes found in temperate climates; for inſtance, the one caught ſome years ago in Sologne, which was entirely white, except a ſingle black feather in the middle of its wings †; whether it had migrated from the northern countries to France, after having undergone this change, or was bred in France, and the change of colour owing to ſome accidental cauſe. We muſt ſay the ſame of the white Magpies that have ſometimes been ſeen in Italy ‡.

Wormius remarks, that the head of his white Magpie was ſmooth and bare, becauſe he probably ſaw it in the moulting ſeaſon; which confirms what I have ſaid with regard to the common Magpies.

Willughby ſaw, in the king of England's collection, Magpies of a brown or ruſty colour, which may be eſteemed a ſecond variety of the ordinary ſort. [A]

* "A white Magpie was hatched in the town of Comarn, in the Ruſſian palatinate. . . . . Five Magpies of the ſame colour were ſeen near Viaſka. In Volhinia, not far from the townſhip of Olika, another was found." RZACYNZKI, *Actuarium*, p. 412.

† Salerne.

‡ GERINI, *Storia degli Uccelli*, tom. ii. p. 41.

[A] Specific character of the Magpie, *Corvus-Pica*, LINN.: "It is variegated with black and white, its tail wedge-ſhaped." Its length is eighteen inches, its breadth only twenty-four. It is found alſo in the iſland of Madeira.

FOREIGN BIRDS,

RELATED TO THE MAGPIE.

I.

The SENEGAL MAGPIE.

IT is ſomewhat leſs than ours; but its wings, being proportionably longer, are nearly of the ſame extent; its tail, on the contrary, is ſhorter, though of the ſame ſhape. The bill, the feet, and the nails, are black, as in the common Magpie, but the plumage is very different. It has not a particle of white, and all the colours are dull; the head, the neck, the back, and the breaſt, are black, with violet reflections; the quills of the tail and the great quills of the wings are brown. All the reſt is blackiſh, with different degrees of intenſity.

II.

The JAMAICA MAGPIE.

This bird weighs only ſix ounces, and is about a third ſmaller than the common Magpie, which

which it reſembles in its bill, its feet, and its tail.

The plumage of the male is black, with purple reflections; that of the female is brown, darker on the back and all the upper ſide of the body, and lighter under the belly.

They build their neſt on the branches of trees. They are found in every part of the iſland, but are moſt numerous at a diſtance from the ſcene of buſtle. After breeding, they quit their concealments, and in autumn they ſpread over the ſettlements in ſuch prodigious multitudes, as ſometimes to darken the air. They fly thus in flocks for miles, and wherever they alight, they occaſion conſiderable damage to the planters. In winter, their reſource is to crowd to the barn-doors. Such facts would lead us to ſuppoſe that they are frugivorous; but they have a ſtrong ſmell, and their fleſh is rank and coarſe, and ſeldom eaten.

It follows from what I have ſaid, that this bird differs from our Magpie, not only in its mode of feeding, in its ſize, and in its plumage, but is beſides diſtinguiſhed by its being able to continue long on wing, by its aſſociating in numerous flocks, and by the rankneſs of its fleſh. The difference of ſex is attended with a ſtill greater in the colours. In ſhort, if we add that the common Magpie could not traverſe the immenſe ocean which ſeparates the two continents, and could not ſupport the in-

tenſe

tenſe cold of a northern paſſage, we may conclude the American Magpies to be analogous to ours, and their repreſentatives in the new world, but not derived from the ſame common ſtock.

The teſquizana * of Mexico ſeems to bear a great reſemblance to this Jamaica Magpie; ſince, according to Fernandez, its tail is very long, and its ſize is inferior to that of the ſtare; its plumage is of a gloſſy black; it flies in numerous flocks, which are deſtructive to the cultivated fields where they alight; it breeds in the ſpring, and its fleſh is tough and rank. In a word, this bird might be conſidered as a ſort of ſtare or jackdaw; but if we except the plumage, a jackdaw with a long tail reſembles much a Magpie.

It is quite different with reſpect to the iſana † of Fernandez, which Briſſon confounds with the Jamaica Magpie. The bill, indeed, the feet, and the plumage, have the ſame colours in both; but the iſana is larger, and its bill is twice as long: beſides, it prefers the coldeſt parts of Mexico, and in its inſtincts, its habits, and its cry, it reſembles the ſtare. It would be difficult, I imagine, to trace theſe characters in

* I have formed this word by contraction for the Mexican *Tequixquiacazanatl*. Fernandez calls it alſo *Stare of Salt Marſhes*, and the Spaniards, *Tordo* (thruſh). This bird has a plaintive ſong. FERNANDEZ, *Hiſt. Avium Novæ Hiſpaniæ*, cap. xxxiv.

† *Id.* cap. xxxii. He calls it *Izanatl*; others, *Yxtlaolzanatl*.

 the

the Jamaica Magpie of Catefby; and, if it muft be referred to the fame genus, it ought at leaft to be formed into a feparate fpecies. The authority of Fernandez, the only naturalift who has had an opportunity of viewing the bird, is furely of more weight than the artificial claffification of a fyftem-maker; and that experienced obferver fays, that it bore a ftronger analogy to the ftare than to the Magpie. However we may be deceived in a fubject of this kind, where our information is drawn from imperfect defcriptions and inaccurate figures; I fhall add, that the ifana has a fort of jeering note, common to moft of the birds termed American *Magpies*.

III.

The MAGPIE of the ANTILLES*.

Corvus Caribæus, Gmel.
Galgalus Antillarum, Briff.
Pica Cauda Indica, Ray.
The Perfian Pie, Will.
The Carribæan Crow, Lath.

Briffon has claffed this bird with the rollers, for no other reafon that I can difcover, except that in Aldrovandus's figure the noftrils are naked,

* See *L'Hiftorie generale des Antilles*, tom. i. p. 258.—*Aldrovandus*, tom. i. p. 788.

naked, which Briſſon reckons one of the characters of the roller: but, 1, we cannot with certainty infer from a figure, which Briſſon himſelf conſiders as inaccurate, a property ſo minute that it would eſcape the notice of a deſigner. 2. To this we may oppoſe a remarkable character, which could not be overlooked, viz. the long quills in the middle of the tail, which Briſſon conſiders as belonging excluſively to the Magpie. 3. The Magpie of the Antilles reſembles ours in its cry, in its confident diſpoſition, in its neſtling on trees, in its ſauntering by the margin of ſtreams, and in the coarſeneſs of its fleſh *: in ſhort, if we muſt rank it with the moſt analogous European birds, it ought to be placed among the Magpies.

It differs, however, by the exceſſive length of the two middle quills of the tail, which ſtretch ſix or eight inches beyond the lateral ones †; its colours are alſo different, the bill and

* *Hiſt. des Antilles.* The Magpie alſo walks by the ſides of water, ſince it ſometimes catches crabs, as we have already ſaid.

† I do not mean the ſingular property aſcribed to it by Aldrovandus, that there are only eight quills in the tail: that naturaliſt counted them only in a coloured figure, a very uncertain method of judging. Father Dutertre, indeed, repeats the ſame thing; but moſt probably he copies Aldrovandus, with whoſe *ornithology* he was well acquainted, ſince he quotes it in the next page. Beſides, he was accuſtomed to make his deſcriptions from memory, which requires aſſiſtance: and, laſtly, the Magpie of the Antilles is perhaps the only one in which he mentions the number of the tail-quills.

feet

feet being red; the neck blue, with a white collar; the head of the ſame blue, tinged with a white ſpot, ſtreaked with black, which extends from the origin of the upper mandible to the junction of the neck; the back of a tawny colour, the rump yellow, the two long quills of the tail ſtriped with blue and white, thoſe of the wing mixed with green and blue, and the under ſide of the body white.

Upon comparing Father du Tertre's deſcription of the Magpie of the Antilles, with that of Aldrovandus's long-tailed Magpie of India, we cannot doubt but they were formed from a bird of the ſame ſpecies, and conſequently it is an American bird, as we are aſſured by Father du Tertre, who ſaw it at Guadaloupe, and not a native of Japan, as Aldrovandus aſſerts from a very uncertain tradition *; unleſs we ſuppoſe that it had penetrated towards the north, and thus ſpread through both continents.

* "The emperor of Japan tranſmitted this moſt beautiful bird, ſome years ſince, to his holineſs the Pope, as a remarkable preſent, as I have learned from the marquis Facchinetto, who ſaid that it was received by his uncle Innocent IX." ALDROVANDUS.

The

IV.

The HOCISANA*.

Corvus Mexicanus, Gmel.
Pica Mexicana Major, Briſſ.
Le Criard, Pernetty's Voy.
The Mexican Crow, Lath.

Though Fernandez calls this bird a great ſtare, we may, from his own account, refer it to the genus of Magpies; for he tells us, that it would be exactly like the common jackdaw, if it were ſomewhat ſmaller, its tail and nails ſhorter, and its plumage of a purer black, and not mixed with blue. But a long tail is the property, not of the ſtare, but of the Magpie, and what diſcriminates it the moſt in its external appearance from the jackdaw. With regard to the other characters which ſeparate the Hociſana from the jackdaw, they are as much foreign to the ſtare as to the Magpie, if not more ſo.

This bird courts the reſidence of man; it is as familiar as the Magpie, chatters like it, and has a ſhrill cry; its fleſh is black, and very well taſted †.

* FERNANDEZ, cap. xxxiii. The Mexican name is *Hocitzanatl*; it is alſo called *Caxcaxtototl*.

† Specific character: "Entirely of a bluiſh black."

V.

The VARDIOLE*.

Seba has given this bird the appellation of the *Bird of Paradiſe*, as he has to almoſt all foreign birds with long tails. In this reſpect the Vardiole was entitled to the name, ſince its tail is double the extreme length of its body. But this tail is not formed as in the bird of paradiſe, for its quill feathers are furniſhed with vanes through their whole length, beſides many other differences.

White is the prevailing colour in this bird: we muſt only except the head and neck, which are black, with very bright purple reflections; the feet, which are of a light red; the wings, whoſe quills have black vanes, and the two middle ones of the tail, which ſtretch much beyond the reſt, and which are marked with black along their ſhaft, from the baſe to half of their length.

The eyes of the Vardiole are lively, and encircled with white; the baſe of the upper mandible is ſhaded with little hair-like black feathers, that meet behind and cover the noſtrils; its wings are ſhort, and extend not beyond the origin of its tail. So far it reſembles the Magpie; but it differs on account of the ſhortneſs of its feet, which are only the half in proportion, a cir-

* It is called *Waygeboe*, or *Wardioe*.

a circumſtance attended with other differences in its figure and port.

It is found in the iſland of Papoe, according to Seba, whoſe deſcription is the only original one, and comprehends all that is known about this bird.

VI.

The ZANOE*.

Corvus-Zanahoe, Gmel.
Pica Mexicana Minor, Briſſ.
The Leſſer Mexican Crow, Lath.

Fernandez compares this Mexican bird to the common Magpie, for its ſize, for the length of its tail, for the perfection of its ſenſes, for its talents for ſpeaking, and for its proneneſs to ſteal whatever pleaſes its fancy. He adds, that its plumage is entirely black, except on the neck and head, where we can perceive a fulvous tinge †.

* The Mexican name is *Tſanahoei*.

† Specific character: "Blackiſh, its head and neck ſomewhat "fulvous, its tail long."

The JAY*.

Le Geai, Buff.
Corvus Glandarius, Linn. and Gmel.
Garrulus, Briſſ.
Pica Glandaria, Geſner, Aldrov. and Ray.

WHAT we have ſaid with regard to the inſtinct of the Magpie, we hold almoſt entirely in reſpect to the Jay; it will be ſufficient, therefore, to notice the characteriſtic differences.

The principal one is the blue ſpot, or rather mail formed by various ſhades of blue, with which each of its wings is decorated, and which ſuffices to diſtinguiſh it, at leaſt, from all the other European birds. It has alſo on its forehead a tuft of ſmall feathers, black, blue, and white: in general its feathers have all a ſoft and ſilky feel, and it can at pleaſure raiſe and depreſs thoſe on its head. It is a fourth part leſs than the Magpie; its tail is ſhorter, and its

* In Greek, according to Belon, Μαλακοκρανευς: in modern Greek, Καρακαξα: in Latin, *Garrulus*: in Spaniſh, *Gayo*, *Cayo*: in Italian, *Ghiandaia*, *Gaza Verla*, *Berta*, *Bertina*, *Baretino*: in German, *Hăher*, *Hătzler*, *Baum Hatzel* (Tree-Jay), *Eichen-heber* (Oak-Jay), *Nuſs-heber* (Nut-Jay), *Nuſs-hecker* (Nut-Hedger), *Jăck*, *Broe-kexter*, *Marggraff*, *Marcolfus*: in Swiſs, *Herren-vogel*: in Poliſh, *Soyka*: in Swediſh, *Net-Skrika*, *Allon-Skrika*, *Korn-Skrika*.

N.º 64

THE JAY.

wings longer in proportion, but notwithſtanding it can ſcarcely fly better *.

The male is diſtinguiſhed by the bulk of his head and the luſtre of his colours †. The old ones differ alſo from the young in their plumage, and hence the various inconſiſtent deſcriptions ‡.

The Jays are of a petulant diſpoſition; they have keen ſenſations and briſk movements, and in their frequent burſts of rage they hurry into danger, and often entangle their head between two branches, and die thus ſuſpended in the air §. When they feel reſtraint, their violence exceeds bounds; and hence, in a cage, they entirely loſe their beauty, by the continual rumpling, wearing, and breaking of their feathers.

Their ordinary cry is harſh and frequent; they are alſo fond of imitating other birds whoſe notes are not more agreeable, ſuch as the keſtril, the tawny owl, &c. ‖ If they perceive in the wood a fox or other ravenous animal, they give

* Belon. † Olina.

‡ Willoughby ſays, that in the Jay deſcribed by Aldrovandus, no tranſverſe ſpots appear on the tail. Its feet are gray, according to Belon; they are brown, verging on fleſh-colour, according to Briſſon, and our own obſervations.

§ Gesner. This inſtinct makes thoſe battles credible, which are ſaid to be fought between armies of Jays and thoſe of Magpies.

‖ Friſch.

a cera certain ſhrill ſcream, to alarm their companions; they quickly aſſemble, preſuming that they ſhall be formidable by their numbers, or at leaſt by their noiſe *. This inſtinct, which the Jays have, of ſummoning their force, together with their violent antipathy to the brown owl, ſuggeſt ſeveral ways of enſnaring them, and the ſport is commonly very ſucceſsful †; for they are more petulant than the magpies, and by no means ſo ſuſpicious or crafty; nor is their natural cry ſo various, though they have great flexibility of throat, and have a turn for imitating all the cries and ſounds of animals which they habitually hear, and even human diſcourſe. The word *Richard* they can the moſt eaſily imitate. The have alſo, like the magpie, and all the family of the daws, crows, and ravens, the habit of burying their ſuperfluous proviſions ‡, and of ſtealing whatever they can obtain. But they cannot always recognize the ſpot where they have buried their treaſure, or, like all miſers, they are more actuated by the fear of encroaching on their ſtock, than by the deſire of uſing it; ſo that in the ſucceeding ſpring, the acorns and nuts that were concealed, perhaps forgotten, germinate in the earth, and their tender leaves diſcover the uſeleſs heap, though too late, to the frugal ſordid hoarders.

* Friſch. † Belon. ‡ *Id.*

The

The Jays breed in woods remote from human dwellings, preferring the moſt branchy oaks, whoſe trunks are entwined with ivy *: but they are not ſo artful and cautious as the magpies in building their neſts. I have received many of theſe in the month of May; they were hollow hemiſpheres, formed with ſmall interwoven roots, open above, without any ſoft lining, and without any exterior defence. I always found them to contain four or five eggs; others ſay that the number is five or ſix. They were ſmaller than pigeons eggs, gray, with more or leſs of a greeniſh hue, and with ſmall ſpots faintly marked.

The young undergo their moulting in July: they keep company with their parents till the ſucceeding ſpring †, when they ſeparate, to form new pairs. By this time the blue plate on their wings, which appears very early, has attained its higheſt beauty.

In the domeſtic condition, to which they eaſily conform themſelves, they become accuſtomed to all ſorts of food, and live in this way eight or ten years ‡. In the ſtate of freedom, they feed not only on acorns and filberts, but on cheſnuts, peas, beans, ſorbs, gooſberries, cherries, raſberries, &c. They alſo prey on the young of other birds, when they can ſurpriſe them in the neſt during the abſence of the pa-

* Olina. † Britiſh Zoology. ‡ Olina, Friſch.

 rents;

rents; and ſometimes they venture to attack the old ones, when they diſcover them entangled in ſnares; and in this caſe they advance with their uſual imprudence, and are often caught themſelves *. Their fleſh, though not delicate, is eatable, particularly if it be boiled firſt, and afterwards roaſted.

In Jays, the firſt phalanx of the outer toe is in each foot connected with that of the middle toe; the inſide of the mouth is black; the tongue of the ſame colour, forked, thin, membranous, and almoſt tranſparent; the gall-bladder is oblong; the ſtomach not ſo thick, and lined with weaker muſcles than the gizzard in the granivorous tribe. Their gullet muſt be very wide, for they ſwallow acorns, filberts, and even cheſnuts entire, like the ring-doves †; I know, however, that they never thus treat the flower-cup of a pink, though they are exceedingly fond of the ſeed which it contains. I have often admired their procedure: if a pink be thrown to them, they ſeize it greedily, and if others be offered, they continue to ſnatch them till their bill can hold no more. When they want to eat theſe, they lay aſide all the reſt but one, hold it with the right foot, and pluck off the petals one by one, keeping a watchful eye all the time, and caſting a glance on every ſide: at laſt, when the ſeed appears, they de-

* Friſch. Britiſh Zoology. † Belon.

vour it greedily, and again begin to pluck a ſecond flower.

This bird is found in Sweden, Scotland, England, Germany, and Italy; and is, I believe, a native of every country in Europe, and even of the correſponding climates of Aſia.

Pliny mentions a kind of Jays or magpies with five toes, which learned to ſpeak better than the reſt *. This is not more wonderful than that there ſhould exiſt hens with five toes, eſpecially as the Jays become more familiar and domeſtic than hens; and we know well, that all animals which live with man, and feed richly, are ſubject to exuberance of growth. The phalanges of the toes might be multiplied in ſome individuals beyond the uſual number; a deviation which has been aſcribed too generally to every ſpecies †.

But another variety, more generally known in this ſpecies, is the White Jay. It has the blue mark on the wings ‡, but is diſtinguiſhed from the common Jay by the almoſt univerſal whiteneſs of its plumage, which extends even to its bill and nails, and by the red colour of its eyes, a property obſerved in ſo many other white animals. But we muſt not imagine that this white complexion is entirely pure; it is often ſhaded with a yellowiſh tinge of various intenſity. In a ſubject which I examined, the

* Lib. x. 42. † Aldrovandus. ‡ Gerini.

coverts of the wings were the whiteſt; its feet alſo ſeemed to be more ſlender than thoſe of the common Jay. [A]

[A] Specific character of the Jay, *Corvus Glandarius*, LINN.: " The coverts of its wings are ſky-blue, with tranſverſe white " and black lines, its body variegated ferruginous." It weighs between ſix and ſeven ounces, its length is thirteen inches, breadth twenty and a half. It is eſteemed one of the moſt beautiful of the Britiſh birds.

M

FOREIGN BIRDS,

WHICH ARE RELATED TO THE JAY.

I.

The RED-BILLED JAY of CHINA.

Corvus Erythrorynchos, Gmel.
The Red-billed Jay, Lath.

THIS new kind of Jay has been juſt introduced into France. Its red bill is the more remarkable, as the whole of the fore-part of the head, the neck, and even the breaſt, is of a fine velvet black. The hind part of its head and neck is of a ſoft gray, which mixes in ſmall ſpots on the crown, with the black of the fore-part; the upper ſide of the body is brown, the under whitiſh. But to form a clear idea of the colours, we muſt ſuppoſe a violet tint ſpread over them all, except the black, deeper on the wings, fainter on the back, and ſtill more dilute under the belly. The tail is tapered, and the wings exceed not one-third of its length, and each of its quills is marked with three colours, viz. a light violet at its origin, black at its middle, and white at its extremity;

 but

but the violet is more extenſive than the black, and that ſtill more than the white.

The feet are red, like the bill; the nails whitiſh at their origin, and brown near the point, and are, beſides, very long and hooked.

This Jay is ſomewhat larger than ours, and may be only a variety ariſing from the influence of climate *.

* Specific character: "The body duſky above, and whitiſh "below; the tail wedge-ſhaped; the tail-quills dilute violet at "the baſe, black in the middle, and white at the tips."

II.

The PERUVIAN JAY.

Le Geai du Perou, Buff.
Corvus Peruvianus, Gmel.

The plumage of this bird is of ſingular beauty; it conſiſts of an aſſemblage of the fineſt colours, ſometimes melting with inimitable art, and ſometimes forming a contraſt which heightens the effect. The delicate green which prevails in the upper part of its body, extends on the one ſide over the ſix mid-quills of the tail, and on the other it advances, paſſing by inſenſible ſhades, and receiving, at the ſame time, a bluiſh tint, to join a ſort of white crown on the head. The baſe of the bill is ſurrounded with a fine blue, which appears again behind the eye, and in in the ſpace below it. A kind of black velvet, which covers the throat and all the fore-part of the neck, is contraſted at its upper margin with the fine blue colour, and at its lower to the jonquil yellow which is ſpread over the breaſt, the belly, and the three lateral quills on each ſide of the tail. The tail is tapered, and more ſo than the Siberian Jay.

Nothing is known with regard to the qualities of this bird, which has never been ſeen in Europe.

III.

The BROWN CANADA JAY, or CINEREOUS CROW. *Lath.*

Corvus Canadenſis, Linn. and Gmel.
Garrulus Canadenſis Fuſcus, Briſſ.
The Cinereous Crow, Penn, and Lath.

If it were poſſible to ſuppoſe that the Jay could migrate into America, I ſhould be inclined to ſuppoſe that this is a variety of our European ſpecies; for it has the appearance and the port, and alſo thoſe ſoft ſilky feathers which are conceived to belong peculiarly to the Jay. It is diſtinguiſhed only by its inferior ſize, by the colours of its plumage, and by the length and ſhape of its tail, which is tapered. Such ſlight differences might be aſcribed to climate;

 but

but our Jay is unable to traverſe the intervening ocean. Till, therefore, we receive a fuller account of the habits of the Brown Jay of Canada, we ſhall conſider it as one of the foreign ſpecies the moſt analogous to our Jay.

The upper ſide of the body is of a brown colour; the under ſide, and alſo the crown of the head, the throat, and the fore-part of the neck, are of a dirty white, which alſo appears at the extremity of the tail and wings. In the individual which I obſerved, the bill and the legs were of a deep brown, the under ſide of the body of a deeper brown, and the lower mandible broader than in the figure: laſtly, the feathers on the throat, jutting forward, formed a ſort of barbil *.

* Specific character: "It is duſky, the front yellowiſh, the "under-ſide of the body and the tips of the tail-quills white." It inhabits the northern and weſtern parts of America, breeds early in the ſpring, and builds its neſt with ſticks and graſs in the pines. It lays two, and rarely three eggs, which are blue. The young are quite black. They fly in pairs, the male and female being preciſely alike. They ſtore up berries in hollow trees; yet they are avaricious, and ſo bold as to viſit the huts of the natives, and pilfer whatever they can ſnatch, even ſalted meat. They are ſeldom ſeen in the month of January, unleſs near dwellings.

IV. The

IV.

The SIBERIAN JAY.

Corvus Sibiricus, Gmel.

The points of analogy between this new ſpecies and our Jay conſiſt in a certain family likeneſs, and that the ſhape of the bill and feet, and the poſition of the noſtrils, are nearly the ſame; and alſo that the Siberian Jay has, like ours, narrow feathers on its head, which it can raiſe at pleaſure as a creſt. The diſcriminating properties are theſe: it is ſmaller, its tail is tapered, and the colours of its plumage are very different. Its hiſtory is totally unknown.

V.

The WHITE COIF, or CAYENNE JAY.

Corvus Cayanus, Linn. and Gmel.
Garrulus Cayanenſis, Briſſ.

It is nearly of the ſize of the common Jay, only it is taller, its bill ſhorter, its tail and wings proportionally longer, which gives it a ſprightlier air.

There are alſo other differences, chiefly in the plumage; gray, white, black, and the different

ferent ſhades of violet, conſtitute all the variety of its colours. The gray appears on the bill, the legs, and the nails; the black on the front, the ſides of the head, and the throat; the white round the eyes, on the crown of the head, and on the nape as far as the origin of the neck, and alſo over all the lower part of the body; the violet lighter on the back and wings, and deeper on the tail, which is tipped with white, and compoſed of twelve quills, of which the two middle ones are rather longer than thoſe towards the ſide.

The ſmall black feathers on its front are ſhort, and ſcarce flexible; part of them project over the noſtrils, and the reſt are reflected, ſo as to form a ſort of ruffled creſt *.

* Specific character: "It is ſomewhat violet, white below, its "neck and front black, its tail white at the tip." It is thirteen inches long.

VI.

The GARLU, or the YELLOW-BELLIED JAY of CAYENNE.

Corvus Flavus, Gmel.

This alſo is a native of Cayenne; but of all the Jays it is the one which has the ſhorteſt wings; we ſhould therefore be the fartheſt from ſuſpect-

ſuſpecting that it croſſed the Atlantic, eſpecially as it can ſubſiſt only in warm climates. Its feet are ſhort and ſlender. I can add nothing with reſpect to its colours, but what the ſight of the figure will ſuggeſt; and with reſpect to its habits, we are totally ignorant. We know not even whether, like the other Jays, it can erect the crown feathers. It is a new ſpecies *.

* Specific character: " Above, it is duſky-greeniſh; below, yellow; its chin and eye-lids white; its wings and tail of a duſky-bluſh colour." It is nine inches long.

VII.

The BLUE JAY of NORTH-AMERICA.

Corvus Criſtatus, Linn. and Gmel.
Garrulus Canadenſis Cæruleus, Briſſ.
Pica Glandaria Criſtata, Klein.
The Blue Jay, Cateſby, Edw. Penn. and Lath.

This bird is noted for the fine blue colour of its plumage, which, with a ſlight intermixture of white, black, and purple, is ſpread over all the upper part of its body, from the crown of the head to the extremity of the tail.

Its throat is white, with a tint of red; under it is a kind of black gorget, and ſtill lower a reddiſh zone, which melts by degrees into the gray and white that predominate in the lower part of the body. The feathers on the crown of of the head are long, and the bird raiſes them at pleaſure like a creſt, which is larger and more beautiful than in our Jay: this is terminated on the front by a kind of black fillet, which, ſtretching on both ſides over a white ground as far as the nape, joins the branches of the gorget. This fillet is divided from the bottom of the upper mandible by a white line formed by the ſmall feathers which cover the noſtrils.

The tail is almoſt as long as the bird itſelf, and conſiſts of twelve ſtaged quills.

Cateſby remarks, that the American Jay has the ſame petulance in its actions as the common Jay; that its notes are leſs diſagreeable, and that the female is diſtinguiſhed from the male by its duller colours. Admitting this, Cateſby's figure muſt repreſent a female, and that of Edwards a male; but the age of the bird muſt alſo affect the vivacity and perfection of its colours.

This Jay is brought from Carolina and Canada; and in thoſe countries it muſt be very common, for many are ſent to Europe *.

* Specific character: "The coverts of the wings are marked "by black tranſverſe lines, its body is cœrulean, its collar black." It is twelve inches long. It feeds on fruits and berries, of which it generally waſtes more than it conſumes. It lays, in the month of May, five or ſix eggs of duſky olive, with ferruginous ſpots. It remains in the country the whole year. It is well known to ſailors by the name of *Blue Bird*, and frequently brought to Britain from Virginia and the Carolinas.

No. 65

THE NUT CRACKER.

The NUTCRACKER*.

Le Caiſſe-Noix, Buff.
Corvus Caryocatactes, Linn. and Gmel.
Nucifraga, Briſſ.
Caryocatactes, Geſner, Ray, and Will.
Merula Saxatilis, Aldrov.

THIS bird is diſtinguiſhed from the jays and magpies by the ſhape of its bill, which is ſtraighter, blunter, and compoſed of two unequal pieces. Its inſtinct is alſo different, for it prefers the reſidence of high mountains, and its diſpoſition is not ſo much tinctured with cunning and ſuſpicion. However, it is cloſely related to theſe two ſpecies of birds; and moſt authors not fettered by their ſyſtems, have ranged it with the jays and magpies, and even with the jackdaws †, which, it is well known, bear a great analogy to the magpies; but it is aſſerted

* This bird was unknown to the Greeks, tho' Geſner has formed a compound Greek name Καρυοκαταχτης, from *καρυα*, a *nut*, and *κτεινω*, *to kill*: in Latin it is called *Nucifraga*, *Oſſifragus*; and by ſome *Turda Saxatilis*, *Pica Abietum Guttata* (Stone-Thruſh, or ſpeckled Pine-Magpie): in Turkiſh, *Gargå*: in German, *Nuſsbretſcher*, *Nuſkraehe* (Nut-Crow), *Tannen heyer* (Fir-Jay), *Steinheyer*, *Wald-ſtarl* (Wood-Stare), *Turkiſcher-holſt-ſchreyer* (the Turkiſh Foreſt-brawler): in Poliſh, *Klcſk*, *Grabuluſk*: in Ruſſian, *Koſtobryz*: in French, *Pie Grivelée*.

† Geſner, Turner, Klein, Willughby, Linnæus, Friſch.

asserted that it chatters more than any of these.

Klein distinguishes two varieties of the Nutcracker; the one, speckled like the stare, has a strong angular bill, a long forked tongue, as in all the magpies; the other is of inferior size, and its bill (for he says nothing of the plumage) is more slender and rounder, composed of two unequal mandibles, the upper of which is the longer, and its tongue divided deeply, very short, and almost lost in the throat*.

According to the same author, these two birds eat hazel-nuts; but the former breaks them, and the latter pierces them: they feed also on acorns, wild berries, the kernels of pine-tops, which they pluck dextrously, and even insects. And lastly, like the jays, the magpies, and the jackdaws, they conceal what they cannot consume.

Besides the brilliancy of the plumage, the Nutcracker is remarkable for the triangular white spots which are spread over its whole body, except the head. These spots are smaller on the

* According to Willughby, the tongue seems not capable of reaching farther than the corners of the mouth, while the bill is closed; because in that situation the cavity of the palate, which usually corresponds to the tongue, is then filled by a protuberant ridge of the lower jaw, which here fits this cavity. He adds, that the bottom of the palate, and the sides of the chaps, are roughened with little points.

upper

upper part, and broader on the breaſt; their effect is the greater, as they are contraſted with the brown ground.

Theſe birds are moſt attached, as I have obſerved above, to mountainous ſituations. They are common in Auvergne, Savoy, Lorraine, Franche-Compté, Switzerland, the Bergamaſque, in Auſtria in the mountains which are covered with foreſts of pines. They alſo occur in Sweden, though only in the ſouthern parts of that country *. The people in Germany call them Turkey birds, Italian birds, African birds; which language means no more than that they are foreign †.

Though the Nutcrackers are not birds of paſſage, they fly ſometimes from the mountains to the plains. Friſch ſays, that flocks of them are often obſerved to accompany other birds into different parts of Germany, eſpecially where there are pine foreſts. But in 1754, great flights of them entered France, particularly Burgundy, where there are few pines; they were ſo fatigued on their arrival, that they ſuffered themſelves to be caught by the hand ‡. One was

* "It inhabits Smoland, and rarely occurs elſewhere." *Fauna Suecica.*—Gerini remarks that it is never ſeen in Tuſcany.

† Friſch.

‡ A ſkilful ornithologiſt of the town of Sarbourg, (Dr. Lottinger, who is well acquainted with the birds of Lorraine, and to whom I am indebted for many facts relating to their inſtincts, their habits,

was killed in the month of October that same year at Mostyn in Flintshire, which was supposed to have come from Germany. We may remark, that that year was exceedingly arid and hot, which must have dried up most of the springs, and have much affected those fruits on which the Nutcrackers usually feed. Besides, as on their arrival they seemed to be famished, and were caught by all sorts of baits, it is probable that they were constrained to abandon their retreats for want of subsistence.

One of the reasons, it is said, why the Nutcrackers do not settle and breed in the inviting climates, is the perpetual war waged against them by the proprietors of the woods, for the injuries which they commit on the large trees, by piercing the trunks, like the wood-peckers *. Part of them is soon destroyed, and the rest is forced to seek an asylum in the desert unprotected forests.

habits, and their migrations), informs me, that in the same year (1754) flights so numerous of Nutcrackers passed into Lorraine, that the woods and the fields were filled with them. Their stay lasted the whole month of October, and hunger had so much enfeebled them, that they were knocked down with sticks. The same observer adds, that these birds appeared again in 1763, but in smaller numbers; that their passage is always in autumn, and that six or nine years commonly intervene between their visits. This must be restricted to Lorraine; for in France, especially in Burgundy, the Nutcrackers appear much seldomer.

* Salerne.

Nor

Nor is this the only circumſtance in which they reſemble the Woodpeckers; they neſtle, like them, in the holes of trees, which, perhaps, they themſelves have formed; for the middle quills of the tail are alſo worn near the end *, which ſhews that they, as well as the woodpeckers, clamber upon trees. In ſhort, Nature ſeems to have placed the Nutcrackers between the Woodpeckers and the Jays; and it is ſingular, that Willughby has given them this preciſe arrangement in his Ornithology, though his deſcription ſuggeſts no relation between theſe ſpecies.

The iris is of a hazel-colour; the bill, the feet, and the nails black; the noſtrils round, ſhaded with whitiſh feathers, ſtraight, ſtiff, and projecting; the feathers of the wing and tail are blackiſh, without ſpots, but only terminated for the moſt part with white; though there are ſome varieties in the different individuals, and in the different deſcriptions, which ſeems to confirm the opinion of Klein with regard to the two races or varieties, which he admits into the ſpecies of the Nutcrackers.

We cannot find, in writers of natural hiſtory, any details with regard to their laying, their incubation, the training of their young, the duration of their life, &c. for they haunt

* Linnæus.

inacceſſible ſpots, where they enjoy undiſturbed ſafety and felicity. [A]

[A] Specific characterof the Nutcracker, *Corvus-Caryocatactes*, LINN.:—" It is duſky, dotted with white, its wings and tail black; the tail-quills white at the tip, the middle ones worn at the tip." It is thirteen inches long. It inhabits Europe and the north of Aſia, but very ſeldom appears in Great Britain.

M

The ROLLERS.

Les Rolliers, Buff.

IF we regard the European Roller as the type of the genus, and reſt its diſtinctive character, not upon one or two ſuperficial qualities, but upon the general combination of its properties, we ſhall be obliged to make conſiderable changes in the enumeration given by Briſſon.

On this principle, which appears to be well founded, I reduce, 1. The European Roller and the Shaga-Rag of Barbary, mentioned by Dr. Shaw, to the ſame ſpecies. 2. I range together the Abyſſinian and the Senegal Roller, with which Briſſon ſeems not to have been acquainted. 3. I claſs together the Roller of Mindanao; that of Angola, which Briſſon makes his twelfth and thirteenth Rollers; and that of Goa, which Briſſon does not mention. 4. I exclude from the genus of Rollers the fifth ſpecies of Briſſon, or the Chineſe Roller, becauſe it is a different bird, and is much more like the Cayenne *Grivert*, with which I ſhall claſs it: I ſhall place both of them, under the common name of *Rolle*, before the Rollers, be-

 cauſe

caufe they appear to form the intermediate ſhade between the Jays and the Rollers. 5. I transfer the Roller of the Antilles to the Jays, which is the ſixth ſpecies of Briſſon. 6. I leave among the birds of prey the *Ytzquauhtli*, of which Briſſon has made his ſeventh ſpecies of Roller, by the name of *the Roller of New Spain*, the hiſtory of which has been given after the Eagles. In fact, according to Fernandez, who is the original author, and even according to Seba, who copies him, it is really a bird of prey, devouring hares and rabbits, and conſequently is very different from the Rollers. Fernandez ſubjoins, that it is proper for falconry, and that its bulk is equal to that of a ram. 7. I omit alſo the *Hoxetot*, or Yellow Roller of Briſſon, which I have ranged after the magpies, as being more related to that kind than to any other. Laſtly, I exclude the *Ococolin* of Fernandez, for the reaſons already ſtated in the article of the quails; nor can I admit the *Ococolin* of Seba, which is very different from that of Fernandez, though it bears the ſame name; for it is of the ſize of a crow, its bill is thick and ſhort, its toes and nails very long, its eyes encircled with red *papillæ*, &c. In ſhort, after this reduction, and the addition of the new ſpecies or varieties which have been hitherto unknown, the genus will conſiſt of two ſpecies of *Rolles*, and ſeven of Rollers with their varieties.

The

The CHINESE ROLLE.

Coracias Sinensis, Gmel.
Galgalus Sinensis, Briss.
The Chinese Roller, Lath.

This bird has wide nostrils like the Rollers, and a bill resembling theirs; but are these characters sufficient to justify its classification with the Rollers? or are these not counterbalanced by more numerous and more important differences? Its feet are longer, its wings shorter, and consist of a smaller number of quills, and these differently proportioned *; its tail is tapered, and its crest is precisely like that of the blue Canada Jay. These circumstances, but particularly the length of its wings, have induced me to assign it a place between the Jays and the Rollers †.

* In the Chinese *Rolle*, the wing consists of eighteen quills, of which the first is very short, and the fifth longer than the rest, as in the Jay; whereas the wing of the Roller includes twenty-three quills, of which the second is the longest.

† Specific character:—"It is green: below, yellowish-"white; the tail wedge-shaped; the tip white." It is eleven inches and a half long.

The GRIVERT, or CAYENNE ROLLE.

Coracias Cayanensis, Gmel.
The Cayenne Roller, Lath.

This bird ought not to be separated from the preceding, which it is entirely like, except because it is smaller, and the colours of its plumage different. With regard to the instincts and habits of these birds, we can draw no comparison, though the resemblance in their exterior properties seems to denote a radical connection*.

* Specific character :—" It is of a dusky green ; below, dirty " white ; the eye-brows white ; the upper part of the throat " striated both ways with black ; the tail wedge-shaped." It is nine inches long.

The GARRULOUS ROLLER†.

Le Rollier d'Europe, Buff.
Coracias-Garrula, Linn. and Gmel.
Galgulus, Briss.
Coracias-Cærulea, Gerini.
Garrulus-Cæruleus, Frisch.

The names of *Strasburg Jay*, *Sea-Magpie*, *Birch-Magpie*, and *German Parrot*, which this bird

† Gesner was told that the German name *Roller* was expressive of its cry ; Schwenckfeld says the same of *Rache*. One of them must

N.° 66

THE GARRULOUS ROLLER.

bird has received in different countries, have been applied at random from popular and ſuperficial analogies. We need only view the bird, or even a good coloured figure of it, to be convinced that it is not a parrot, though there is a mixture of green and blue in its plumage; and a cloſer examination will inform us that it is neither a magpie nor a jay, though it chatters inceſſantly like theſe birds *. Its appearance and port are different; its bill is not ſo thick; its legs much ſhorter in proportion, ſhorter even than the mid-toe; its wings longer, and its tail entirely of a different ſhape, the two outer quills projecting more than half an inch (at leaſt in ſome individuals) beyond the ſix intermediate ones, which are all equal in length. It has alſo a kind of wart behind the eye, and the eye itſelf is ſurrounded with a ring of yellow naked ſkin †.

The appellation of Straſburg Jay is ſtill more abſurd; for M. Hermann, profeſſor of medicine and natural hiſtory in that city, writes me,

muſt be miſtaken, and I am inclined to think that it is Geſner; for the name *Rache*, adopted by Schwenckfeld, is more analogous with thoſe given to this bird in different countries, and which are probably derived from its cry. In German, *Galgen-Regel, Halk-Regel, Gals-Kregel, Racher*: in Poliſh, *Kraſka*: in Swediſh, *Spanſk-Kraſka*. It has alſo the following names in Germany: *Heiden-Elſter, Kugel-Elſter, Mandel-Krae, Deutſcher-Papagey, Birk-Heher*, (i. e. Heath-Magpie, Ball-Magpie, Almond-Crow, German Popinjay, Birch-Jay.)

* Aldrovandus. † Edwards.

" The Rollers are ſo rare here, that ſcarcely " three or four ſtragglers are ſeen in the courſe " of twenty years." One of theſe had been ſent to Geſner, who, not being acquainted with the fact, denominated it the *Straſburg Jay*.

Beſides, it is a bird of paſſage, and performs its migrations regularly once a-year, in the months of May and September *; yet it is not ſo common as the magpie or the jay. It is found in Sweden † and in Africa ‡; but we muſt not ſuppoſe it ſettled in the intermediate regions. It is unknown in many parts of Germany §, France, and Switzerland ||, &c. We may therefore conclude that, in its paſſage, it moves only in a narrow zone, from Smoland and Scania to Africa. There are even points enow given to mark nearly its tract through Saxony, Franconia, Suabia, Bavaria, Tirol, Italy ¶, Sicily **, and laſtly, the iſland of Malta ††, which is a ſort of general rendezvous for all

* Extract of a Letter from the Commander Godeheu of Riville, on the Migration of Birds, tom. iii. *Memoires preſentés à 'Academie Royale des Sciences*, p. 82.

† *Fauna Suecica*, No. 73.

‡ Shaw's Travels.

§ Friſch.

|| " It was caught with us in the middle of Auguſt 1561, and " not known." GESNER *de Avibus*.

¶ " I remember to have ſeen it once at Bologna." *Id.*

** " We ſaw them for ſale on the ſtalls at Meſſina in Sicily." WILLUGHBY.

†† " We ſaw them expoſed for ſale in the market of Malta." *Id.* Alſo Commander Godeheu's letter.

the

the birds that croſs the Mediterranean. The one deſcribed by Edwards was killed on the rock of Gibraltar, whence it could wing its lofty * courſe to the African ſhore. It is alſo ſeen ſometimes in the vicinity of Straſburg, as we have already noticed, and even in Lorraine, and in the heart of France †; but theſe are probably young ones, which ſtray from the main body.

The Roller is more wild than the jay or the magpie: it ſettles in the thickeſt and the moſt ſolitary woods; nor, as far as I know, has it ever been tamed or taught to ſpeak ‡. Its plumage is beautiful; it has an aſſemblage of the fineſt ſhades of blue and green, mixed with white, and heightened by the contraſt of duſky colours §. But a good figure is ſuperior to any deſcription. The young do not aſſume the delicate azure till the ſecond year; whereas the jays are decorated with their moſt beautiful feathers before they leave the neſt.

* Geſner.

† Brisson. M Lottinger informs me, that in Lorraine theſe birds paſs more ſeldom than the Nutcrackers, and in ſmaller numbers. He adds, that they are never ſeen but in autumn, no more than the Nutcrackers; and that in 1771 one was wounded in the neighbourhood of Sarrebourg, which, notwithſtanding, lived thirteen or fourteen days without ſuſtenance.

‡ Schwenckfeld.

§ *Linnæus* is the only one who ſays that its back is blood-coloured. *Fauna Suecica*, No. 73.—Was the ſubject that he deſcribed different from all thoſe deſcribed by other naturaliſts?

The

The Rollers build, when it is in their power, on birches, and it is only when they cannot find theſe that they lodge in other trees *. But in countries where wood is ſcarce, as in the iſland of Malta and in Africa, they form their neſt, it is ſaid, on the ground †. If this be a fact, it would follow, that the inſtincts of animals can be modified by ſituation, climate, &c.

Klein ſays, that contrary to what happens in other birds, the young Rollers void their excrements in the neſt ‡; and this circumſtance has perhaps given riſe to the notion that this bird beſmears its neſt with human ordure, as has been alleged of the hoopoe §; but this is inconſiſtent with its lonely ſylvan haunt.

Theſe birds are often ſeen in company with the wood-peckers and crows, in the tilled

* Friſch.

† "A ſportſman," ſays M. Godeheu, in a letter which I have already quoted, "aſſured me, that in the month of June he ſaw "one of theſe birds iſſuing from a bank of earth, where was a "hole as large as the hand; and that having dug the ſpot in the "direction of the hole, which went horizontally, he found, at the "depth of a foot or thereabouts, a neſt made of ſtraw and thorns, "in which were two eggs." This account of the ſportſman, which would be doubtful if it were ſingle, ſeems confirmed by that of Dr. Shaw, who, ſpeaking of the bird known in Africa under the name of *Shaga-Rag*, ſays, that it makes its neſt on the brinks of rivers. Notwithſtanding, I am much afraid that there is ſome miſtake, and that the King-fiſher was taken for the Roller, on account of the reſemblance of its colours.

‡ *Ordo Avium*, p. 62.

§ Schwenckfeld.

grounds

grounds which are in the vicinity of their forefts. They pick up the fmall feeds, roots, and worms which the plough throws to the furface, and even the grain that is lately fown. When this fupply fails them, they have recourfe to wild berries, caterpillars, grafshoppers, and even frogs *. Schwenckfeld adds, that they fometimes devour carrion; but this muft be during winter, and only in cafes of abfolute want †; for they are in general regarded as not carnivorous, and Schwenckfeld himfelf remarks that they are very fat in autumn, and then are good eating ‡, which can hardly be faid of birds that feed on garbage.

The Roller has long narrow noftrils placed obliquely on the bill near its bafe, and open; the tongue is black, not forked, but ragged at the tip, and terminated towards the root by two forked appendices, one on each fide; the palate is green, the gullet yellow, the ventricle of a faffron colour, the inteftines about a foot long, and the *cæca* twenty-feven lines. The wings extend twenty-two inches, each confifting of twenty quills, or, according to others, of twenty-three, the fecond of which is the longeft of all. Laftly, it is obferved that

* Klein, Willughby, Schwenckfeld, Linnæus.

† If they rake among garbage in fummer, it muft be for infects.

‡ Frifch compares their flefh to that of the ring-dove.

wherever

wherever theſe quills are black on the outſide, they are blue beneath.

Aldrovandus, who ſeems to have been well acquainted with theſe birds, and who lived in a country which they inhabit, aſſerts that the female differs much from the male, its bill being thicker, and its head, neck, breaſt, and belly of a cheſnut colour, bordering on aſh-gray, while the correſponding parts in the male are of the colour of the beryl, with different reflections of a duller green. I ſuſpect that the two long outſide quills of the tail, and the warts behind the eyes, which appear only in ſome individuals, are the attributes of the male, as the ſpur in the gallinaceous tribe, the long tail in the peacocks, &c. [A]

* Willughby, Schwenckfeld, Briſſon.

[A] Specific character of the Garrulous-Roller, *Coracias-Garrula*, LINN.:—"It is ſky-blue, its back red, its tail-quills black." Its eggs are of a pale green, with numerous dull ſpots, and of the bulk of a pigeon's. It is ſeldom or never ſeen in Great Britain.

M

VARIETIES of the ROLLER.

Dr. Shaw mentions, in his Travels, a bird of Barbary, called by the Arabs *Shaga-Rag*, which is of the bulk and ſhape of the jay, but with a ſmaller bill and ſhorter feet.

The

The upper part of the body of this bird is brown; the head, neck, and belly of a light green, and on the wings, as well as on the tail, are ſpots of a deep blue. Dr. Shaw adds, that it makes its neſt on the banks of rivers, and that its cry is ſhrill.

This ſhort deſcription agrees ſo well with our Roller, that we cannot doubt but the Shaga-Rag belongs to the ſame ſpecies; and the reſemblance which the name bears to moſt of the German appellations of the Roller, derived from its voice, adds to the probability *.

* Mr. Latham conjectures that the *Shaga-Rag* is the ſame with the Variety of the Abyſſinian Roller, afterwards deſcribed.

FOREIGN BIRDS,

WHICH ARE RELATED TO THE ROLLER.

I.

The ABYSSINIAN ROLLER.

Coracias Abyssinica, Gmel.

THIS bird is, in its plumage, much like the European Roller; only its colours are more lively and brilliant, which must be ascribed to the influence of a drier and hotter climate. On the other hand, it resembles the Angola Roller, by the length of the two side feathers of its tail, which project five inches beyond the rest. In short, this bird seems to occupy a place between the European and Angola Rollers. The point of its upper mandible is very hooked. It is entirely a new species.

VARIETY *of the* ABYSSINIAN ROLLER.

We may consider the Senegal Roller as a variety of that of Abyssinia. The chief difference between them is, that in the Abyssinian bird

bird the orange colour of the back does not extend, as in that of Senegal, ſo far as the neck and the hind part of the head: a difference which would not be ſufficient to conſtitute two diſtinct ſpecies, eſpecially as they belong to nearly the ſame climate, as the two lateral quills are double the length of the intermediate ones, as in both the wings are ſhorter than thoſe of the European Roller; and laſtly, as they are alike in the ſhades, the luſtre, and the diſtribution of their colours *.

* This is the *Coracias Senegalenſis* of Gmelin, the *Swallow-tailed Indian Roller* of Edwards, and the *Senegal Roller* of Latham.

II.

The ANGOLA ROLLER, or the MINDANAO ROLLER.

Theſe two Rollers reſemble each other ſo exactly, that it is impoſſible to ſeparate them. That of Angola is diſtinguiſhed from the other only by the length of the exterior quills of its tail, which is double that of the intermediate ones, and by ſlight variations of colour. But differences ſo minute may be the effect of age, of ſex, or even of moulting; and the inſpection of our figures, nay, the deſcriptions of

 Briſſon,

Briſſon, who makes two ſpecies of them, will confirm our conjecture of the identity of the two ſpecies. They are both nearly of the bulk of the European Roller, have the ſame general ſhape, its bill ſomewhat hooked, its naked noſtrils, its ſhort legs, its long toes, its long wings, and even the colours of its plumage, though differently diſtributed: they are always blue, green, and brown, which are ſometimes diſtinct, ſometimes mixed, melted together, forming many intermediate ſhades, and having various reflections. The bluiſh green, or ſea green, is however ſpread on the crown of the head; the brown, more or leſs intenſe, and more or leſs greeniſh, covers all the fore-part of the body, with ſome tints of violet on the throat; and the blue, the green, and all the ſhades which ariſe from their mixture, appear on the rump, the tail, the wings, and the belly: only the Mindanao Roller has under its breaſt a kind of orange tincture, which is not found in that of Angola.

To this opinion it will be objected, perhaps, that the kingdom of Angola is at a great diſtance from Bengal, and ſtill farther from the Philippines. But is it impoſſible, or is it not natural, that theſe birds ſhould be ſpread through the different parts of the ſame continent, or the neighbouring iſlands, which are connected with it perhaps by the continuation of the ſame chain,

chain, especially in climates so nearly alike? Besides, we cannot always expect the most scrupulous exactness in those who import the productions of foreign countries; and the intercourse of European vessels with the various regions of the globe is so extensive and multiplied, that a bird found in the East Indies, might have been carried to Guinea, and afterwards imported as a native of Africa. Admitting this, if we ascribe the slight differences between the Roller of Mindanao and that of Angola to the effect of age, we must reckon the latter the older; or if we impute them to the distinction of sex, we must consider it as the male: for we know that in the Rollers, the fine colours of the feathers do not appear till the second year; and it is a general principle, that in all birds, the male, when it differs from the female, is distinguished by an exuberance of growth, or a superior richness of plumage *.

* Specific character of the Angola Roller, which is the *Coracias-Caudata* of Linnæus, and the *Long-tailed Roller* of Latham:—" It is somewhat fulvous; below, cœrulean; the neck " striated below with pale violet; the outmost quills of the tail " very long."

Specific character of the Mindanao Roller, which is the *Coracias-Bengalensis* of Gmelin; the *Bengal Pie*, or *Jay*, of Albin; and the *Bengal Roller* of Latham:—" It is somewhat fulvous; " below, cœrulean; the neck striated beneath with pale violet; " the tail entire."

 VARIETY

VARIETY *of the* ANGOLA *and* MINDANAO ROLLERS.

The Royal Cabinet has lately received from Goa a new Roller, which is very like that of Mindanao. It differs only by its ſize, and by a ſort of collar, like wine-lees in colour, which graſps only the hind part of the neck, a little under the head. It has not, any more than the Angola Roller, the orange cincture of the Mindanao Roller; but if in this reſpect it differs from the latter, it is ſo much the more allied to the former, which is certainly of the ſame ſpecies.

III.

The ROLLER of the INDIES.

Coracias Orientalis, Gmel.
Galgulus Indicus, Briſſ.
The Oriental Roller, Lath.

This Roller, which is the fourth of Briſſon, differs leſs from the preceding in the nature of its colours, which are always blue, green, brown, &c. than in the order of their diſtribution; but in general its plumage is more duſky, its bill is alſo broader at the baſe, more hooked, and

No. 67

THE MADAGASCAR ROLLER,

and of a yellow colour: laſtly, of all the Rollers it has the longeſt wings.

M. Sonerat has lately ſent to the Royal Cabinet a bird, which is almoſt in every reſpect like the Indian Roller; only its bill is ſtill broader, and for this reaſon it has received the epithet of *large-toad-mouthed:* but that appellation would better ſuit the Goat-ſucker *.

* Specific character:—" It is green, its throat ſtriated with " cœrulean; its tail-quills black at the tip. It is of the bulk of " the jay, being ten inches and a half long."

IV.

The MADAGASCAR ROLLER.

Coracias Madagaſcarienſis, Gmel.

This ſpecies differs from all the preceding in ſeveral properties: its bill is thicker at the baſe, its eyes are larger, its wings and tail longer, though the exterior pupils of the latter do not project beyond the reſt: laſtly, the plumage is of an uniform purple-brown, excepting only that the bill is yellow, the largeſt quills of the wings black, the lower belly of a light blue, the tail of the ſame colour, edged at its extremity with a bar of three ſhades, viz. purple, light blue, and dark purple approaching to black. It has all the other characters which belong to the Rollers; ſhort feet, the edges of the upper mandible ſcalloped near the point, the ſmall feathers which reflect from its baſe, and the naked noſtrils, &c.

V.

The MEXICAN ROLLER.

Coracias Mexicanus, Gmel.
Galgulus Mexicanus, Briſſ.

This is the Mexican Black-bird of Seba, which Briſſon makes his eighth Roller. It would require the inſpection of it to fix its true ſpecies; for this would be difficult, from the ſhort notice given by Seba, who is here the original author. I place it among the Rollers, becauſe I know of no reaſon to exclude it; I therefore follow the opinion of Briſſon, till more perfect information confirm or deſtroy the temporary arrangement. The colours are different from thoſe which are common in the Rollers. The upper part of the body is of a dull gray, mixed with a rufous tint, and the under of a light gray, with ſome marks of fire-colour *.

* Specific character:—"It is of a gray-rufous; below, and on "the wings, of a dilute gray, mixed with flame-colour." It is much larger than a thruſh.

VI. The

No. 68

THE GREATER PARADISE.

VI.

The PARADISE ROLLER.

Oriolus Aureus, Linn. and Gmel.
Paradiſea Aurea, Lath.
Ićtericus Indicus, Lath.
The Golden Bird of Paradiſe, Edw.

I place this bird between the Rollers and the Birds of Paradiſe, as forming the ſhade which connećts theſe two kinds, becauſe it ſeems to have the ſhape of the former, and to reſemble the latter by its ſmallneſs, and the ſituation of the eyes under and very near the junćtion of the mandibles, and by a ſort of natural velvet which covers the throat and part of the head. Beſides, the two long quills of the tail, which ſometimes occur in the European Roller, and which are much longer in that of Angola, is another analogical charaćter that connećts the genus of the Roller with that of the Bird of Paradiſe.

The upper part of the body of this bird is of a vivid and brilliant orange, the under of a fine yellow; it has no black but under the throat, on part of the ſhoulders, and on the quills of the tail. The feathers which cover the hind part of the neck are long, narrow, flexible, and recline on each ſide over the lateral parts of the neck and breaſt.

The feet and legs had been torn from the ſubject deſcribed and deſigned by Edwards, as if it had been a real Bird of Paradiſe ; and this circumſtance probably led that naturaliſt to refer it to that genus, though it has none of the principal characters. The quills of the wings were wanting, though thoſe of the tail were complete ; they were, as I have ſaid, twelve in number, and terminated with yellow. Edwards ſuſpects that the quills of the wing are alſo black, whether becauſe they are of the ſame colour with thoſe of the tail, or that they were wanting in the individual which he obſerved ; for dealers in birds, in drying the ſpecimens, pluck all the feathers which are of a bad colour, to increaſe the beauty of the plumage *.

* Specific character :—" It is of a fulvous-yellow ; its bridle, " the upper part of its throat, the primary coverts, and the tips " of the tail-quills, black." It is eight inches long.

The GREATER BIRD of PARADISE*.

L' Oiseau de Paradis, Buff.
Paradisea Apoda, Linn. and Gmel.
Manucodiata, Briss.
Paradisea Avis, Clusius, Seba, Wormius, &c.

THIS species is more famous for the fictitious and imaginary qualities ascribed to it, than for any real and remarkable properties. The name of the *Bird of Paradise* commonly suggests the idea of a bird which has no feet; which flies constantly, even in its sleep, or at most suspends itself but for a few moments from the branches of trees, by means of the long filaments of its tail †; which copulates in its flight, like certain insects, and lays and hatches in a way unexampled in nature ‡; which lives only on vapours and dews, and which has the cavity

* In Latin, *Apis Indica*, *Avis Dei*, *Parvus Pavo*, *Pavo Indicus*, *Manucodiata*, which the Italians have adopted, *Manucodiata Rex*, *Manucodiata Longa*, *Hippomanucodiata*, *Hirundo Ternatensis*: in German, *Luft-Vogel* (Sky-Bird), *Paradiss-Vogel*: in Portuguese, *Passaros de Sol* (Sparrow of the Sun).

† Acosta.

‡ To give an air of probability to the relation, the male, it is alleged, has on its back a cavity, where the female deposits her eggs, and hatches them by means of a corresponding cavity in

vity of its abdomen entirely filled with fat, inſtead of ſtomach and inteſtines *, (which would be quite ſuperfluous, ſince it eats nothing, and therefore needs not to digeſt or to void:) in ſhort, which has no exiſtence but motion, no element but air, where it is ſupported as long as it retains breath, as fiſh are buoyed up in water, and which never touches the ground till after death †.

This monſtrous heap of abſurdities is only a chain of conſequences juſtly drawn from a radical error, that the Bird of Paradiſe has no legs, though it is furniſhed with even pretty large ones ‡.

The fact § is, that the Indian merchants, who trade with the feathers of this bird, or the fowlers

in her *abdomen*; and that the ſitter might maintain her poſture, they entwine themſelves with their long filaments. Others have ſaid, that they neſtle in the terreſtrial paradiſe, and hence their name. See *Muſæum Wormianum*, p. 294.

* Aldrovandus.

† The people of India ſay, that they are always found with their bills pitched into the ground. *Navigations aux Terres Auſtrales*, tom. ii. p. 232. In fact, their bill muſt neceſſarily fall foremoſt.

‡ Barrere, who ſeems on this head to ſpeak only from conjecture, aſſerts, that the Birds of Paradiſe have legs ſo ſhort, and ſo thickly clothed with feathers to the toes, that one ſhould ſuppoſe them to have none at all. It is thus that, trying to explain one miſtake, he falls into another.

§ The inhabitants of the Arou iſlands believe that theſe birds are hatched with legs, but apt to loſe them, either from diſeaſe or old age If this were true, it would at once explain and excuſe the error. See the obſervations of J. Otto Helbigius, *Collect. Acad.*

fowlers who ſell them, are accuſtomed, whether for the ſake of preſerving and tranſporting the ſpecimens with more eaſe, or perhaps of countenancing an error which is favourable to their intereſt, to dry the bird with its feathers, after having previouſly ſeparated the thighs and extracted the entrails. This practice has been ſo long continued, as to have ſtrengthened the prejudice to ſuch a degree, that thoſe who firſt aſſerted the truth were, as uſual, regarded as unworthy of credit *.

The fable, that the Bird of Paradiſe continually flies, derived an appearance of probability from the conſideration of the quantity of feathers with which it is furniſhed; for beſides thoſe common to other birds, it has many long feathers, which riſe on each ſide between the wing and the thigh, and which, extending much beyond the true tail, and mingling with it, form a ſort of falſe tail, which many obſervers have miſtaken. Theſe *ſubalar* feathers are what the naturaliſts term *decompoſed*; they are very light themſelves, and form a bunch

Acad. partie Etrang. tom. iii. p. 448.) If what Olaus Wormius (*Muſæum*, p. 295.) aſſerts were a fact, that each of the toes of this bird has three articulations, this ſingularity would be ſtill greater; for in almoſt all birds, the number of joints is different in each toe, the hind one having two, including that of the nail, and of the fore-toes, the inner having three, the mid-one four, and the outer five.

* "Antonius Pigafetta falſely aſcribes to their legs a palm of "length." *Aldrovandus*, tom. i. p. 807.

almoſt

almoſt devoid of weight, and aërial; they will therefore increaſe the apparent bulk of the bird *, diminiſh its ſpecific gravity, and thus aſſiſt in ſupporting it in the air. But if the wind be contrary, the abundance of plumage will rather obſtruct its motion; accordingly it is obſerved, that the bird of Paradiſe avoids the bluſtering gales †, and commonly ſettles in countries the leaſt ſubject to them.

Theſe feathers are of the number of forty or fifty on each ſide, of unequal lengths; the greater part ſpread under the true tail, and others lie over it, without concealing it; for their texture is delicately ſlender, and almoſt tranſparent, which is very difficult to repreſent in a figure.

Theſe feathers are highly eſteemed in India, and much ſought after. It is not more than a century ſince they were employed in Europe for the ſame purpoſes as thoſe of the Oſtrich; and, indeed, their lightneſs and brilliancy make them elegant ornaments. But the prieſts of Aſia aſcribe to them miraculous virtues, which give them a new value in the eyes of the vulgar, and have procured the bird the appellation of the *Bird of God*.

* It is ſaid to appear as large as a pigeon, though it exceeds not the bulk of a blackbird.

† The Arous conſiſt of five iſlands, and theſe birds inhabit only the middle ones; they never appear in the others, becauſe, being naturally weak, they cannot withſtand high winds. HELBIGIUS.

Next

Next to this, the moſt remarkable property of the Bird of Paradiſe is thoſe two long filaments which take their riſe above the true tail, and extend more than a foot beyond the falſe tail, formed by the ſubalar feathers. These, indeed, are real filaments only at their middle; for at their origin and their termination, they are furniſhed with webs of the ordinary breadth. In the females the extremities are narrower, which, according to Briſſon, is the only diſtinction between it and the male *.

The head and throat are covered with a ſort of velvet, formed by ſmall erect feathers, which are ſhort, ſtiff, and cloſe; thoſe of the breaſt and back are longer, but always ſilky and ſoft to the feel. They are all of different colours, which vary according to the poſition and the light in which they are viewed.

The head is very ſmall in proportion to the body; the eyes ſtill ſmaller, and placed very near the opening of the bill. Cluſius reckons only ten quills in the tail; but this aſſertion was certainly not founded on the examination of a living ſubject, and it is doubtful whether the plumage of a bird brought from ſo great a diſtance be entire, eſpecially as it is ſubject to an annual moulting, which laſts ſeveral months. During that time, which happens in

* The inhabitants of the country ſay, that the females are ſmaller than the males, according to J. Otto Helbigias.

in the rainy ſeaſon, it lives concealed; but, in the beginning of Auguſt, after hatching, its feathers are reſtored, and in the months of September and October, in which calm weather prevails, it flies in flocks, like the Stares in Europe *.

This beautiful bird is not much diffuſed: it is almoſt entirely confined to that part of Aſia which produces the ſpiceries, and eſpecially the iſlands of Arou. It is known alſo in the part of New Guinea oppoſite to theſe iſlands; but the name which it there receives, *Burung-Arou*, ſeems to indicate its natal ſoil.

Since warm regions of ſpices alone are proper for the Bird of Paradiſe, it probably ſubſiſts on ſome aromatic productions †; at leaſt it does not live ſolely on dew. J. Otto Helbigius, who travelled into India, tells us, that it feeds on red berries, which grow on a very tall tree. Linnæus ſays, that it ſubſiſts on large butterflies ‡; and Bontius, that it ſometimes preys on ſmall birds. Its ordinary haunt is the woods, where it perches on the trees, and the Indians watch it in ſlender huts, which they

* Helbigius.

† Tavernier remarks, that the Paradiſe Bird is very fond of nutmegs, and that it reſorts to eat them in the ſeaſon; that it paſſes in flocks, like thoſe which we obſerve of the thruſhes in the time of vintage, and that they are intoxicated by the nutmegs, and drop down. *Voyage des Indes*, tom. iii. p. 369.

‡ Syſtema Naturæ, *Edit*. x. p. 110.

attach

attach to the branches, and ſhoot it with their arrows of reeds *. It flies like the ſwallow, whence it has been called the *Ternate-ſwallow* †; though others ſay, that its ſhape, indeed, reſembles the ſwallow, but that it flies higher, and always ſoars in the aërial regions ‡.

Though Marcgrave ranges it among the birds of Brazil, there is no reaſon to ſuppoſe that it exiſts in America; at leaſt no European veſſels have ever imported it from thence. Beſides, that naturaliſt does not, as uſual, mention the name which it receives in the language of the Brazilians, and a bird, clothed in ſuch delicate ſwelling plumage, could not traverſe the wide expanſe of ocean which divides the equatorial parts of the two continents.

The ancients ſeem to have been totally unacquainted with the Bird of Paradiſe: no mention is ever made of its rich decorations. Belon pretends that it was the phœnix of antiquity; but his opinion is founded on the fabulous qualities of both §. The phœnix, too, appeared in

* Some open the belly with a knife, as ſoon as they drop, and, having detached the entrails with a part of the fleſh, they introduce into the cavity a red-hot iron; after which they dry the bird in the chimney, and ſell it for a low price to the merchants. HELBIGIUS.

† Bontius.

‡ *Navig. aux Terres Auſtr.* tom. ii. p. 252.

§ " It has a golden brilliancy about its neck; its other parts " are purple," ſays Pliny, ſpeaking of the Phœnix; then he adds, " no perſon ever ſaw it feed." Lib. x. 2.

in Arabia and Egypt, while the Bird of Paradiſe has remained always attached to the Oriental parts of Aſia, which were very little known to the ancients.

Cluſius mentions, on the authority of ſome mariners, who themſelves learned the fact from report, that there are two kinds of this bird; the one large and beautiful, which inhabits the iſlands of Arou; the other inferior to it in ſize and elegance, which is ſettled in the country of the Papous, next Gilolo *. Helbigius, who heard the ſame in the iſlands of Arou, adds, that the Birds of Paradiſe of New Guinea, or of the Papous, differ from thoſe of Arou, not only in point of ſize, but alſo in the colours of the plumage, which is white and yellowiſh. I ſhould regard theſe authorities as ſuſpicious, and inſufficient to found any general concluſion. The dried ſpecimens indeed, which are brought to Europe, preſent great diverſity of appearance; in ſize, in the number and poſition of the feathers, in the colours of the plumage, &c. But, in ſuch mutilated and imperfect preparations, it is impoſſible to decide what muſt be aſcribed to the effect of age, of ſex, of ſeaſon, of climate, and of other accidental cauſes. Beſides, the Birds of Paradiſe

* J. Otto Helbigius ſpeaks of the ſpecies which is found in New Guinea, as not having in its tail the two long filaments which appear in that of the ſpecies of the Arou iſlands.

being very expenſive articles of commerce, many other birds, with long tails and an elegant plumage, have been paſſed on the credulity of the public, and the legs and thighs pulled off, to conceal the fraud and enhance the price. We have already had an example in the Paradiſe Roller, mentioned by Edwards, on which the honours of mutilation had been conferred. I have myſelf ſeen ſeveral paroquets, promerops, and other birds, which had been thus treated, and many inſtances are to be found in Aldrovandus and Seba: and it is very common to disfigure the real Birds of Paradiſe, with a view to add to their value. I ſhall therefore take notice only of two principal ſpecies of theſe birds, without venturing to vouch for the accuracy of that diviſion till new obſervations illuſtrate the matter *.

* Specific character of the *Paradiſea Apoda* of LINNÆUS:—
"The feathers on the flanks are longer than the body; the two "middle tail-quills long and hairy."

The MANUCODE.

Paradisea Regia, Gmel.
Manucodiata Minor, Briss.
Rex Avium Paradisæarum, Gaza, Seba, Clusius, &c.
The King's Bird, Forrest.
The King Paradise Bird, Lath.

I ADOPT this name from the Indian appellation *Manucodiata*, which signifies *Bird of God*. It is usually called *the King of the Birds of Paradise*; but this appellation is drawn from fabulous accounts. Clusius was informed by the mariners, from a tradition which prevailed in the East, that each of the two species of the Birds of Paradise had its leader, whose imperial mandates were received with submissive obedience by a numerous train of subjects: that his majesty always flew above the flock, and issued orders for inspecting and tasting the springs, where they might drink with safety, &c. * This ridiculous fable is what alone consoles Nieremberg for the loss of the multitude of vulgar opinions which Clusius has erased from the history of birds; and this, by the

* This may allude to the method by which the people of India sometimes take whole flocks of birds, by poisoning the fountains to which they resort and drink.

way,

N.° 69

THE KING PARADISE.

way, may ſerve to fix our idea of that compiler's judgment.

The King Bird of Paradiſe reſembles much the reſt. Like them, his head is ſmall, his eyes ſtill ſmaller, placed near the corner of the opening of the bill; his feet pretty long and firm; the colours of his plumage gloſſy; the two filaments of his tail nearly ſimilar, except that they are ſhorter, and their extremity, which is furniſhed with webs, forms a curl, by rolling into itſelf, and is ornamented with ſpangles, reſembling in miniature thoſe of the peacock *. He alſo has beneath the wing, on each ſide, a bunch of ſeven or eight feathers, which are longer than in moſt birds, but not ſo long as thoſe of the Bird of Paradiſe, and of a different ſhape, for they are edged through their whole extent with webs of adhering filaments. The Manucode is ſmaller, the bill white and long in proportion; the wings are alſo longer, the tail ſhorter, and the noſtrils are covered with feathers.

Cluſius counted only thirteen quills in each wing, and ſeven or eight in the tail; but he did not conſider that in a dried ſpecimen theſe might be complete. The ſame author remarks as a ſingularity, that in ſome the two filaments of the tail croſs each other, though this might

* *Collection Academique*, tom. iii. *Part. Etran.* p. 449.

often happen from accident, confidering their flexibility and their length *.

* Specific character of the *Paradifea Regia* of Linnæus:— "The two middle tail-quills are thread-like, their tip of a "crefcent-fhape, and feathery."

M

The MAGNIFICENT BIRD OF PARADISE †.

Le Magnifique de la Nouvelle Guinée, ou *Le Manucode à Bouquets*, Buff.
Paradifea Magnifica, Gmel.

The two tufts *(bouquets)* which I regard as the diftinctive character of this bird, appear behind the neck and at its origin. The firft confifts of feveral narrow feathers of a yellow colour, marked near the point with a fmall black fpot, and which, inftead of lying flat as ordinary, ftand erect, thofe near the head at right angles, and the fucceeding ones with fmaller inclinations.

Under the firft tuft we perceive a fecond, which is larger, but not fo much raifed, and more reclined: it is compofed of long detached filaments, which fprout from very fhort fhafts, and of which fifteen or twenty join together,

† This bird bears fome relation to the *Manucodiata-Cirrhata* of Aldrovandus. The latter has a fimilar tuft, formed in the fame way of unwebbed feathers, but which appears longer, and its bill and tail are much longer.

forming

forming ſtraw-coloured feathers. Theſe feathers ſeem to be cut ſquare at the end, and make angles, more or leſs acute, with the plane of the ſhoulders.

This ſecond tuft is bounded on the right and left by common feathers, variegated with brown and orange, and is terminated behind by a reddiſh and ſhining brown ſpot, of a triangular ſhape, with the vertex turned towards the tail, and the filaments of the feathers looſe and decompoſed, as in the ſecond tuft.

Another characteriſtic feature of this bird is the two filaments of the tail, which are about a foot long and a line broad, and of a blue colour, changing into a lucid green, and taking their origin above the tail. So far they much reſemble the filaments of the preceding ſpecies, but are of a different form, for they do not end in a point, and are furniſhed with webs on the middle only of the inner ſide.

The middle of the neck and breaſt is marked from the throat by a row of very ſhort feathers, diſplaying a ſeries of ſmall tranſverſe lines, which are alternately of a fine light green, changing into blue, and of a deep duck-green.

Brown is the prevailing colour on the lower belly, the rump, and the tail; ruſty yellow is that of the quills, the wings, and of their coverts; but the quills have more than one brown ſpot at their extremity, at leaſt this is the caſe in the ſpecimen preſerved in the Royal Cabinet; for it may be proper to mention that the long quills of the wings, as well as the feet, have been removed *.

This bird is rather larger than the preceding; its bill is ſimilar, and the feathers of the front extend over the noſtrils, which they partly cover: this is inconſiſtent with the character that has been eſtabliſhed of theſe birds by one of our moſt intelligent ornithologiſts †.

The feathers of the head are ſhort, ſtraight, cloſe, and very ſoft to the touch. They form a ſort of velvet of a changing colour, as in almoſt all the Birds of Paradiſe, and of a browniſh ground. The throat is alſo covered with velvet feathers; but theſe are black, with golden-green reflections. [A]

* I know not whether the individual obſerved by Aldrovandus had the number of wing-quills very complete; but this author ſays that theſe quills were of a blackiſh colour.

† The feathers at the baſe of the bill turned back, and leaving the noſtrils bare. BRISSON.

[A] Specific character of the *Paradiſea Magnifica*:—" It is " ſcarlet above; the upper part of its throat green, with golden, " creſcents; the neck bearing a bunch of yellow feathers."

M

The

The BLACK MANUCODE of NEW GUINEA, called the SUPERB.

Paradiſea Superba, Gmel.
The Superb Paradiſe Bird, Lath.

The predominant colour of the plumage of this bird is a rich velvet black, decorated under the neck with reflections of deep violet. Its head, breaſt, and the hind part of its neck, are brilliant, with the variable ſhades of a fine green; the reſt is entirely black, not even excepting the bill.

I place this bird immediately after the Birds of Paradiſe, though it wants the filaments of the tail; but we may ſuppoſe that moulting, or ſome accidental cauſe, is the reaſon of this defect; for in other reſpects it reſembles theſe birds, not only in its general ſhape, and in that of its bill, but is alſo related by the identity of climate, by the richneſs of its colours, and a certain ſuperabundance or luxuriancy of feathers which is peculiar to the Birds of Paradiſe: for there are two ſmall tufts of black feathers which cover the noſtrils, and two other bunches of the ſame colour, but much longer, and directed to the oppoſite extremity. Theſe riſe on the ſhoulders, and ſpreading more or leſs over the back, but always bent backwards, form a

 ſort

ſort of wings, which extend almoſt to the extremity of the true, when theſe are cloſed.

We muſt add, that theſe feathers are of unequal lengths, and that thoſe of the anterior ſurface of the neck and the ſides of the breaſt are very long and narrow. [A]

[A] Specific character of the *Paradiſea Superba*:—" It is " ſomewhat creſted with a gold-green; below, it is a lively " green; the upper part of the throat violet; its wings black; " its tail blue and ſhining."

M

The SIFILET, or MANUCODE with Six Filaments.

Paradiſea Aurea, Gmel.
The Gold-breaſted Bird of Paradiſe.

If we adopt the filaments as the ſpecific character of the Manucodes, the preſent is entitled to be ranged at their head; for inſtead of two, it has ſix, and of theſe not one riſes on the back, but all of them take their origin from the head, three on each ſide. They are half a foot long, and reflect backwards. They have no webs but at their extremity for the ſpace of ſix lines, and theſe are black and pretty long.

Beſides theſe filaments, this bird has two properties which belong to the Bird of Paradiſe; luxuriancy of feathers and richneſs of colours.

The

The luxuriancy of feathers confifts; 1. In a fort of tuft compofed of ftiff narrow feathers, and which rifes at the bafe of the upper mandible. 2. In the length of the feathers of the belly and of the abdomen, which is four inches or more; one part of thefe feathers, extending directly, conceals the under-fide of the tail, while another part, rifing obliquely on each fide, covers the upper furface of the tail as far as the third of its length, and all of them correfpond to the fubalar feathers of the Bird of Paradife, and of the Manucode.

With regard to the plumage, the moft brilliant colours appear on the neck; behind, it is gold-green and bronze violet; before, topaz-gold reflections, which wanton in all the fhades of green, and derive new luftre from the contraft with the darknefs of the contiguous parts; for the head is black, changing into a deep violet, and the reft of the body is brown, inclining to black, and with reflections of the fame deep violet.

The bill of this bird is nearly the fame as in the Birds of Paradife; the only difference is, that its upper ridge is angular and fharp, while in moft of the other kinds it is rounded.

Nothing can be faid with refpect to the feet and the wings, becaufe they were extirpated in the fubject from which this defcription is drawn; a practice which, as we have remarked, is usual with the Indian hunters or merchants. [A]

[A] Specific character of the *Paradisea Aurea*:—" It is " crusted with black; the top, the cheeks, and the upper part " of its throat are glistening violet; the rest of the throat, the " breast, and the spot on the neck, gold-green."

M

The CALYBE' of NEW GUINEA*.

Paradisea Viridis, Gmel.
The Blue Green Paradise Bird, Lath.

If this bird has not the luxuriant plumage of the Paradise tribe, it has at least the rich colours and the peculiar softness of texture.

Its head is covered with a beautiful blue velvet, changing into green, and exhibiting the reflections of the beryl. The neck is clothed with a longer shag, but which dazzles with the same colours, except that each feather, being of a shining black in the middle, of a green changing into blue only at the edges, there result waving shades, which play still more than those of the head. The back, the rump, the tail and the belly are blue, like polished steel, and with very brilliant reflections.

* The name *Calybé*, or Calybete, was given by Daubenton the younger, to express the chief colour of its plumage, which is a bronzed steel. To the same gentleman we owe the elements of the descriptions of these four new species.

The

The ſmall velvet feathers on its forehead project forwards as far as the noſtrils, which are deeper than in the preceding kinds. The bill is alſo longer and thicker, but it is of the ſame ſhape, and its edges are ſcalloped in the ſame manner near the point. Six quills only are reckoned in the tail, but probably it was not entire.

In the ſubject on which this deſcription is founded, as well as thoſe of the three preceding deſcriptions, a ſtick was paſſed through their whole length, and projected two or three inches out of the bill *. In that ſimple way, and by extirpating the feathers which would ſpoil the effect, the Indians can in an inſtant form an elegant ſort of plume with any ſmall bird which they meet. But the ſpecimens are thus deranged, and their proportions altered. On this account it was difficult to diſcover in the Calybé the inſertion of the wings; inſomuch that credulity might have aſſerted that this bird had neither feet nor wings.

The Calybé differs from the Manucodes more than the preceding: for this reaſon I have ranged it in the laſt place, and beſtowed on it a particular name †.

* They were brought from India by M. Sonnerat, correſpondent of the king's cabinet of natural hiſtory.

† Specific character of the *Paradiſea Viridis*:—"It is ſea-"green; its back, belly, rump, and tail, ſteel-coloured." It is ſixteen inches long.

The OX-PECKER.

Le Pique-Bœuf, Buff.
Buphaga Africana, Linn. and Gmel.
Buphaga, Briss.
The African Beef-eater, Lath.

BRISSON is the firſt who has deſcribed this little bird, which Adanſon brought from Senegal. It is ſcarcely larger than the creſted lark, and its wings extend only fourteen inches. Its plumage has nothing remarkable; in general a grayiſh brown prevails on the upper part of the body, and grayiſh yellow on the under. The bill is not of an invariable colour; in ſome individuals it is entirely brown; in others red at the point, and yellow at the baſe; in all it is nearly of a quadrangular ſhape, and the points of the two mandibles are reflected in a contrary direction. The tail is tapered in ſteps, and a ſingular circumſtance is obſerved, that the twelve quills, of which it conſiſts, are all pointed. Laſtly, the firſt *phalanx* of the exterior toe is cloſely connected to that of the mid-toe.

This bird is very fond of certain worms, or the *larvæ* of inſects, which lodge under the epidermis in oxen. It alights on the backs of theſe animals, and pierces their ſkin with its bill, to extract theſe worms, and hence its name.

N.° 70

THE AFRICAN BEEF-EATER

N.o 71

THE STARE.

The COMMON STARE*.

L'Etourneau, Buff.
Sturnus Vulgaris, Linn. Gmel. Scop. Kram, &c.
Sturnus, Gesner, Belon, Aldrov. Briss. &c.
The Stare, or *Starling*, Will. Ray. Sibb. Alb. Penn, &c.

Few birds are so generally known, especially in the temperate climates, as the Stare; for as it is a constant resident of the district where it settles †, and as it can be trained in the domestic state, its habits have been observed, whether when subjected to restraint, or ranging without controul.

The Blackbird is that, of all the feathered race, which the Stare resembles the most; their

* In Greek ψαρος, whence the name for Granite, ψαρονιον, the spotted surface of that stone resembling the plumage of the Stare; it had also the appellations, Αςραλος, Βαθος, Γολμις or ψολμις: in Latin, *Sturnus* or *Sturnellus*: in Hebrew, *Sarsir*, or *Zezir*: in Arabic, *Alzarazir*, *Zarater*, *Azuri*: in Italian, *Sturno*, *Storno*, *Stornello*: in Portuguese, *Sturnino*: in Spanish, *Estornino*: in German, *Staar*, *Starn*, *Spreche*, *Rinder-Starn* (Ox-Stare): in Flemish, *Spreuve*, *Sprue*: in Polish, *Szpak*, *Spatzck*, *Szpaczieck*, *Skorzek*.

† In the colder countries, however, such as Sweden and Switzerland, it is migratory. "It descends after the middle of summer into the low lands of Scania," says Linnæus, *Fauna Suecica*, p. 70.: "when they leave our country," says Gesner, *de Avibus*, p. 745.

young

young can hardly be diftinguifhed *: but after their characters are developed, the Stare is found to be diftinguifhed by the ftreaks and reflections of its plumage; by the blunter form of its bill, which is broader, and not fcalloped near the point †; and by the greater flatnefs of its head, &c. But another very remarkable difference, and which is derived from a more intimate caufe is, that the fpecies of the Stare is folitary in Europe; whereas the fpecies of the Blackbird are extremely numerous.

There is another circumftance alfo in which thefe birds are analogous; they never change their refidence during the winter. They only feek for thofe fpots in the tract where they are fettled that have the beft afpect, and are in the neighbourhood of fprings ‡; with this difference, however, that the Blackbirds ftill continue to live folitary; whereas the Stares affemble after the breeding feafon, in very numerous flocks: thefe fly in a peculiar manner, which would

* Belon. So exact is the refemblance between the young of thefe two fpecies, that I once knew a law-fuit grounded upon it. The appellant produced a well trained blackbird, and demanded the recompence ftipulated for rearing and educating it; but the defendant infifted, that the young bird which he had committed in charge was a Stare.

† Barrere fays, that the Stare has a quadrangular bill. He muft at leaft allow that the angles are very round.

‡ This has probably led Ariftotle to fay, that the Stare conceals itfelf well in winter.

feem

ſeem to be directed by a ſort of tactics. It is the voice of inſtinct which inceſſantly impels the Stares toward the centre of the battalion, while the rapidity of their motions hurries them beyond it; a ſort of vortex is thus formed, denſer at the middle, and rarer near the verge; and the collective body performs an uniform circular revolution, and at the ſame time continues to make a progreſſive advance. This mode of flying has its advantages and its inconveniencies. The rapacious tribe is diſconcerted by the whirling of the Stares, alarmed by their noiſy cries, and deterred by the appearance of order. But the danger is increaſed of falling a prey to the arts of man: the bird-catcher fixes a packthread beſmeared with bird-lime to each foot, and thus diſcharges one or two Stares; theſe mingle in the flock of their companions, and in their frequent gyrations and rencounters entangle others, and a number of victims, after waſting their efforts, tumble headlong to the ground.

The evening is the time when the Stares aſſemble in the greateſt numbers, to provide more effectually againſt the dangers of the night; which they commonly ſpend among the reeds, whither they haſten about the cloſe of the day, in a noiſy flight *. They chatter much in the

* OLINA. He compares the noiſe of their paſſage through the air to the rattling of hail.

evening

evening and morning, at the forming and difpersing of their forces; are lefs clamorous during the reft of the day, and quite filent during the night.

So attached are the Stares to fociety, that they not only join thofe of their own fpecies, but alfo birds of a different kind. In the fpring, before the breeding feafon, they often affociate with the crows and jackdaws, and even with the red-wings and field-fares, and fometimes with the pigeons.

Their loves commence about the end of March. Violence decides their choice: they continue as noify as ever; their twitter is inceffant; and to fing and toy feem then their fole occupation. The care of the expected progeny fucceeds; but they are not over-anxious in providing for the reception. They often take poffeffion of the neft of a wood-pecker, which often retaliates in its turn. When they would conftruct one for themfelves, they only heap a few dry leaves, fome herbage or mofs, in the hole of a tree or of a wall. In this artlefs bed the female drops five or fix eggs, of a greenifh afh-colour, and covers them for the fpace of eighteen or twenty-one days. Sometimes fhe hatches in dove-cotes, in the roofs of dwelling-houfes, and even in the holes of rocks on the fea-coaft, as in the Ifle of Wight and

in

in other places *. I have ſometimes received, in the month of May, neſts which were pretended to belong to the Stare, and to be found in trees; but as two of them reſemble exactly thoſe of the Thruſh, I ſuſpect that the perſons whom I employed wanted to impoſe on me, unleſs we ſuppoſe that the Stare had diſpoſſeſſed the Thruſh, and occupied its place. In ſome caſes, however, they make their own neſts; a ſkilful obſerver told me, that he has ſeen ſeveral of them on the ſame tree. The young continue long with their mother, which would make me doubt the aſſertion of ſome authors †, that the Stare hatches thrice a-year; except this relate to warm countries, where the progreſs of incubation and of growth is rapid.

The feathers of the Stares are in general long and narrow, as Belon deſcribes; their colour is at firſt of a blackiſh-brown, uniform, and without ſtreaks or reflections. The ſtreaks begin to appear after the firſt moulting, emerging about the end of July, on the lower part of the body, then on the head, and towards the 20th of Auguſt, are ſpread over the upper part of the body I always mean the young Stares, which were hatched in the beginning of May.

I have remarked, that in this firſt moulting, the feathers which ſurround the baſe of the bill,

* Britiſh Zoology.

† "It has two or three hatches annually, each conſiſting of "four or five young." OLINA, *Uccelliera.*

bill, dropped almoſt all at once, ſo that this part was bald during the month of July *, as it happens conſtantly in the rook through the reſt of the year. I alſo obſerved that the bill was almoſt all yellow on the 15th of May; this ſoon changed into a horn colour, and Belon aſſures us, that in time it becomes orange.

In the males, the eyes have a larger ſhare of brown, or it is more uniform †; the ſtreaks of the plumage more diſtinct and yellower; and the dark colour of the feathers which are not ſtreaked is brightened by more vivid reflections, that vary between purple and deep green. Beſides, the male is larger, weighing three ounces and a half. Salerne adds, another diſtinguiſhed character of the ſex is, that the tongue is pointed in the male, but forked in the female. It would appear that Linnæus had ſeen it pointed in ſome individuals, and forked in others ‡. In thoſe which came under my examination, it was forked.

The Stares live on ſnails, worms, and caterpillars; eſpecially on thoſe large caterpillars of

* I know not why Pliny, ſpeaking of the Stares, ſays, "But "theſe loſe not their feathers." Lib. x. 24.

† "The female has a little mail in the pupil of the eyes, "which in the male is entirely black." OLINA.—Willughby ſeems alſo to allude to this ſlough on the eye: "The irides are "hazel, the upper part whiter," where he muſt probably mean the female.

‡ *Lingua Acutâ*. Syſt. Nat. Edit. x.—*Lingua bifidâ*, Fauna Suecica.

fine green, with reddish reflections, which appear, in the month of June, upon the flowers, and chiefly upon the roses. They feed also on wheat, buckwheat, millet, panic, hemp-seed, elder-berries, olives, cherries, raisins, &c. It is pretended that the last is what corrects best the natural bitterness of their flesh, and that cherries are what they are the fondest of*; and these afford an almost infallible bait for weel-nets, which are laid among the reeds, where they retire in the evenings; and in this way an hundred may often be caught in one night: but this diversion lasts no longer than the season of cherries.

They are fond of following oxen and other large cattle as they feed in the meadows, attracted, it is said, by the insects which flutter round them, or by those, perhaps, which swarm in their dung, or in meadows in general. From this habit is derived the German name *Rinder-Staren*. They are also accused of feeding on the carcasses that are exposed on gibbets †; but it is probably in search only of insects. I have raised some of these birds, and have observed, that when bits of raw flesh were offered to

* Schwenckfeld, Salerne, &c. Cardan says, that to sweeter the flesh of Stares, we need only cut off their head as soon as they are killed: Albin directs them to be flayed: others allege, that the mountain Stares are better tasted than those of the plains. But these authors can only mean the young birds, for the flesh of the old ones is always bitter and unpalatable.

† Aldrovandus.

 them,

them, they fixed on the prey with great avidity. If they were preſented with the flower-cup of a pink, containing ſeed already formed, they did not graſp it with their claws, and pluck it like the jay, but ſhook it and ſtruck it againſt the bars in the bottom of the cage, till the grains dropped out. I remarked alſo, that they drank nearly like the gallinaceous tribe, and took great delight in bathing. It is probable that one of thoſe which I raiſed died of cold, in conſequence of bathing too often during the winter.

Theſe birds live ſeven or eight years, or even longer, in the domeſtic ſtate. The wild ones cannot be decoyed by the call, becauſe they regard not the ſcream of the owl. But beſides the contrivance of the limed threads and the weel-nets, which I have already mentioned, a method has been fallen on to take entire families, by fixing to the walls and the trees where they lodge pots of earthen ware of a convenient form, which the birds often prefer to place their neſts in *. Many are alſo caught by the gin and draw-net. In ſome parts of Italy it is common to employ tame weaſels to drag them out of their neſts, or rather their holes; for the artifice of man conſiſts in employing one enſlaved race to extend his dominion over the reſt.

The Stares have the *membrana nictitans*; their noſtrils are half-ſheathed by a membrane; the

* Olina and Schwenckfeld.

legs

legs are of a reddiſh brown *; the outer toe is connected to the mid one as far as the firſt *phalanx*; the hind nail is ſtronger than the reſt; the gizzard is preceded by a dilatation of the *æſophagus*, is a little fleſhy, and contains ſometimes ſmall ſtones. The inteſtinal tube is twenty inches in length, from the one orifice to the other; the gall-bladder is of the ordinary ſize; the *cæca* very ſmall, and placed nearer the anus than is common in birds.

In diſſecting a young Stare, one of thoſe which I had raiſed, I remarked that the contents of the gizzard and of the inteſtines were entirely black, though it had been fed on bread and milk only. This circumſtance denotes an abundance of black bile; and at the ſame time accounts for the bitterneſs of the fleſh of theſe birds, and the uſe which has been made of their excrements in the preparation of coſmetics.

The Stare can be taught to ſpeak either French, German, Latin, Greek †, &c. and to pronounce phraſes of ſome length. Its pliant throat accommodates itſelf to every inflexion and every accent. It can readily articulate the letter R ‡,

* I know not why Willughby ſays that "the legs are feathered "to the toes." I never found this in any of the Stares which I examined.

† "The young Cæſars had a Stare and Nightingales docil in "the Greek and Latin languages, and which made continual pro- "greſs, and aſſiduouſly prattled new phraſes of conſiderable "length." PLINY, lib. x. 42.

‡ Scaliger.

 and

and acquires a ſort of warbling which is much ſuperior to its native ſong *

This bird is ſpread through an extenſive range in the ancient continent. It is found in Sweden, Germany, France, Italy, the Iſle of Malta, the Cape of Good Hope †, and every where nearly the ſame; whereas thoſe American birds which have been called Stares, preſent a great diverſity of appearance. [A]

* *Sturnus piſitat ore, iſitat, piſiſtrat:* It was thus that the Latins expreſſed the notes of the Stare. See the *Author of Philomela.*

† Kolben.

[A] Specific character of the Stare, *Sturnus Vulgaris*, LINN. ——"Its bill is yellowiſh, its body black, with white points." It is near nine inches long, and its alar extent fourteen inches. The male weighs three ounces, the female ſomewhat leſs.

M

VARIETIES *of the* STARE.

Though the Stares retain uniformly the original impreſſion, they are not entirely exempted from the tendency to variety in nature; but the varieties which occur are always ſuperficial, and often confined to individuals. The following have been noticed by authors:

I. The WHITE STARE of Aldrovandus ‡, with fleſh-coloured legs and a reddiſh yellow bill,

‡ *Sturnus Vulgaris.* var. 1. Linn. and Gmel.; *Sturnus Albus*, Briſſ.; *the White Starling*, Will.

bill, as in the common kind after they have grown old. Aldrovandus ſays, that it was taken along with the ordinary Stares; and Rzaczynſki informs us, that in a certain part of Poland * it was uſual to ſee a Black and a White Stare riſing from the ſame neſt. Willughby alſo ſpeaks of two White Stares which were obſerved in Cumberland.

II. The BLACK and WHITE STARE.—To this variety I refer; 1. The White-headed Stare of Aldrovandus †: In this bird, the head, the bill, the neck, the whole of the under part of the body, the coverts of the wings, and the two exterior quills of the tail, were white; the other quills of the tail, and all thoſe of the wings, were as in the ordinary Stare; the white of the head was ſet off by two ſmall black ſpots placed above the eyes, and the white of the under part of the body was variegated with bluiſh ſpots. 2. The Pied Stare of Schwenckfeld, in which the top of the head, the half of the bill next the baſe, the neck, the quills of the wing, and thoſe of the tail, were black, and all the reſt white. 3. The Black-headed Stare, ſeen by Willughby, the reſt of the body entirely white ‡.

* Near Coronovia.

† *Sturnus Vulgaris*, var. 3. Linn. and Gmel.; *Sturnus Leucocephalus*, Briſſ.

‡ *Sturnus Vulgaris*, var. 2. Linn. and Gmel.; *Sturnus Leucomelas*, Briſſ.; *the Black and White Starling*, Will.

III. The GREY CINEREOUS STARE of Aldrovandus *. This author is the only perſon who has ſeen one of that colour, which is nothing but black melted with white. It is eaſy to conceive how theſe varieties might be multiplied from the different diſtribution of the black and white, and from the numerous ſhades of gray, which reſult from the different proportions in which the two original colours enter into the mixture.

* *Sturnus Vulgaris*, var. 4. Linn. and Gmel; *Sturnus Cinereus*, Briſſ.

FOREIGN BIRDS,

RELATED TO THE STARE.

I.

The CAPE STARE, or the PIED STARE.

Sturnus Capensis, Linn. & Gmel.

THIS African bird resembles in its general shape the Common Stare, and the black and white colours of its plumage are distributed as in the Magpie.

Were it not that its bill is thicker and longer than in the European Stare, we might regard it as merely a variety, especially as our Stare is to be met with at the Cape of Good Hope; and this variety would coincide with the one already mentioned, in which the black and white are distributed in large spots. The most remarkable character in this bird is a very large white spot, of a round shape, placed on each side of the head, and which stretches forward to the base of the bill, and inclosing the eye shoots into a sort of appendix, variegated with black, that descends along its neck.

This bird is the same with Edwards's Black and White Indian Starling, Pl. 187.; with Albin's

Contra of Bengal, vol. iii. Pl. 31.; with Brisson's Cape of Good Hope Stare, vol. ii. p. 448.; and even with his ninth tropic bird. He acknowledges this, and rectifies it, p. 54. of the Supplement; and considering the chaos of incomplete description, and of mutilated figures, which disgraces Natural History, he is certainly excusable. To avoid confusion, therefore, it is of the utmost importance to collate the different names which have been bestowed on a bird by different authors, and at different times *.

* Specific character of the *Cape Stare*:—"It is blackish, "the sides of the head and the under part of the body white." The other birds mentioned are considered by Mr. Latham as varieties; but Gmelin is rather disposed to form them into a new species, the *Sturnus Contra*, whose character is, "dusky, with a "spot on the eye, a stripe on the wing, and the belly all white."

II.

The LOUISIANA STARE, or the STOURNE.

Sturnus Ludovicianus, Linn. & Gmel.

I have applied the name of *Stourne*, which is formed from the Latin *Sturnus*, to an American bird, which, though considerably different, is allied to our Stare. The under part of its body is is gray, variegated with brown, and the upper is yellow. The moſt characteriſtic marks of this bird, in reſpect to colour, are, 1. A blackiſh plate intermixed with gray at the lower part of the neck, and riſing diſtinctly out of the yellow ground. 2. Three white bars on its head, which riſe from the baſe of the upper mandible, and extend as far as the *occiput*; the one reſts on the top of the head, while the two others, which are parallel to it, ſtretch on each ſide over the eyes. In general this bird reſembles the European Stare, by the proportions of its wings and tail, and alſo by the diſperſion of theſe colours in ſmall ſpots: its head is likewiſe flat, but its bill is longer.

A correſpondent of the Cabinet informs us, that Louiſiana is much incommoded by clouds of theſe Stares; which would ſhow that their manner of flying reſembles that of the European ſort. But we are not very certain if he means the ſpecies of this article *.

* Specific character of the *Louiſiana Stare*:—"It is ſpotted "with gray and duſky, a line on the head, and the eye-lids "white; the upper part of its throat black."

III. The

III.

The TOLCANA*.

Sturnus Obſcurus, Gmel.
Sturnus Novæ Hiſpaniæ, Briff.
The Brown-head Stare, Lath.

The ſhort account which Fernandez has given of this bird is not only incomplete, but careleſsly drawn up; for though he ſays that the Tolcana is, in ſize and figure, like the Stare, he afterwards adds that it is rather ſmaller. Yet he is the only original author from whom we can obtain information with regard to this bird, and on his evidence Briſſon has ranged it among the Stares. It appears to me, however, that theſe two authors adopt very different characters of the Stare: Briſſon, for inſtance, makes it the diſtinguiſhing feature of the genus, that the bill is ſtraight, blunt, and convex; and Fernandez, ſpeaking of a bird of the *Tzanatl* or *Stare* kind, mentions, that it is ſhort, thick, and rather hooked; and in another place he refers the ſame bird named *Cacalototl* to the genus of the Raven (which is called *Cacalotl* in the Mexican language),

* Formed from the Mexican name *Tolocatzanatl*, which ſignifies Reed-ſtare.

language), and to that of the Stare *. The arrangement of the *Tolcana* is therefore not determined; I have retained its Mexican name, without venturing to pronounce whether it is a Stare or not.

This bird is, like the European Stares, fond of places abounding in rushes and aquatic plants. Its head is brown, and the rest of its plumage black. It has no song, or even cry. In that it partakes of the qualities of many other American birds, which are more remarkable for the richness of their plumage than the sweetness of their warble †.

* "*Cacalotototl*, or Raven Bird, seems to belong to the genus "*Tzanatl* of the Stares."

This bird has, according to Fernandez, a black plumage inclining to blue, the bill entirely black, the iris orange, the tail long, the flesh bad to eat, and no song. It loves temperate and warm countries. From this short statement, it is difficult to decide whether the bird in question be a Raven or a Stare.

† Specific character of the *Sturnus Obscurus*:—" It is black, " its head dusky."

IV.

The CACASTOL ‡.

Sturnus Mexicanus, Gmel.
Cotinga Mexicana, Briss.
The Mexican Stare, Lath.

I here range this bird on the very suspicious authority of Fernandez, and the analogy which its

‡ The Mexican name is *Caxcaxtototl*; it is also called *Hueitzanatl*. *Tzanatl* in the Mexican language corresponds to our Stare.

its name bears in the Mexican language to that of the Stare; nor am I acquainted with any European bird to which I can refer it. Briſſon, who conceives it to be a *Cottinga*, has been obliged, in order to ſupport his preconceptions, to reject from the deſcription of Fernandez, already too ſhort, the words which indicate the lengthened and pointed ſhape of its bill; this ſhape being really related more cloſely to the Stare than to the Cottinga. Beſides, the Caſtol is nearly of the bulk of the Stare; and, like that bird, it has a ſmall head, and is indifferent food; it likewiſe inhabits the temperate and warm regions. It is indeed a bad ſinger, but we have ſeen that the native notes of the European Stare are not very captivating; and, if it were carried into America, we may preſume that its imitation of the harſh muſic of the foreſt would ſoon deſtroy every harmonious modulation *.

* Specific character of the *Sturnus Mexicanus*:—"It is ſky-"blue, variegated with black."

V.

The PIMALOT†.

The broad bill of this bird might lead us to ſuſpect that it is not a Stare; but if what Fernan-

† This word is formed from the Mexican name *Pitzmalotl.*

dez

dez ſays be true, that its habits and inſtincts are the ſame with thoſe of the other Stares, we cannot heſitate to infer that it is of an analogous ſpecies; eſpecially as it generally haunts the coaſts of the South Sea, lodging probably, like the European ſort, among the aquatic plants.—The Pimalot is rather larger.

VI.

The STARE of TERRA MAGELLANICA, or, the WHITE RAY.

Sturnus Milibaris, Gmel.
The Magellanic Stare, Lath.

I have given this laſt name to a bird, brought by M. Bougainville, on account of the white ray which, riſing on each ſide near the junction of the mandibles, bends under the eye and ſtretches along the neck. This white ray is the more remarkable, as it is environed by a deep brown; the dark colours prevail on the upper part of the body, only the wings and their coverts are edged with yellow. The tail is of a full black, forked, and extending not far beyond the wings, which are very long. The under ſide of the body, including the throat, is of a fine crimſon red, ſprinkled with black on the ſides; the anterior

terior part of the wing is alſo of crimſon, and not ſpotted ; and the ſame colour appears round the eyes, and in the ſpace lying between theſe and the bill, which, though blunt as in the Stares, and leſs pointed than that of the Troupiales, may be regarded as neareſt the ſhape of the latter. If we conſider alſo that the White Ray reſembles much the appearance of the Troupiales, we may eſteem it as intermediate between theſe two kinds *.

* Specific character of the *Sturnus Milibaris* :—" It is duſky, " red below."

The TROUPIALES.

THESE birds, as I have juſt obſerved, are nearly related to the European Stares, and often the vulgar and the naturaliſts have confounded them. We may regard them as repreſenting the Stares in the New World; their habits are the ſame, except in the mode of building their neſts.

The American continent is the native region of theſe birds, and of all others that have been claſſed with them; ſuch as the Caſſics, the Baltimores, the Bonanas, &c.; and though ſome are ſaid to belong to the Old World, theſe have really been brought from the New World; for inſtance, probably, the Troupiale of Senegal, called the *Cape-More* *, the Bonana of the Cape of Good Hope, and all the pretended Troupiales of Madras.

I ſhall exclude from the genus of the Troupiales, 1. The four ſpecies brought from Madras, and which Briſſon has borrowed from Ray; becauſe the *law of climate* will not admit the ſuppoſition, and the deſcriptions are not deciſive, and the figures ſo ill executed, that they might as well be taken for magpies, jays, blackbirds, loriots, and gobe-mouches, &c.

* *Oriolus Textor*, Gmel; *the Weaver Oriole*, Lath.

A ſkilful ornithologiſt (Mr. Edwards) is of opinion, that the yellow jay and the chop jay of Petiver, which Briſſon has made his ſixth and his fourth Troupiale, are only the male and female loriot; and that the variegated jay of Madras of the ſame Petiver, which is the fifth Troupiale of Briſſon, is his yellow Indian Stare; and laſtly, that the creſted Troupiale of Madras, which is Briſſon's ſeventh ſpecies, is the ſame bird with the creſted gobe-mouche of the Cape of Good Hope of the ſame Briſſon *.

2. I ſhall exclude the Bengal Troupiale, which is Briſſon's ninth ſpecies, ſince that author has himſelf perceived that it is his ſecond of the Stare.

3. I ſhall exclude the Forked-tail Troupiale, which is the ſixth of Briſſon, and the Thruſh of Seba. All that the latter ſays is, that it is much larger than the thruſh; that its plumage is black, its bill yellow, the under ſurface of its tail white, the upper and its back ſhaded by a light tint of blue; that its tail is long, broad, and forked; and laſtly, that, excepting the difference in the ſhape of its tail, and in its bulk, it is much like the European thruſh. But in all this, I can perceive nothing that relates to a Troupiale; and the figure given by Seba, and which Briſſon rec-

* He adds, that the two long quills of the tail were wanting in theſe two individuals. They muſt either have not yet grown, or have dropped by moulting or ſome other accident.

kons

kons a very bad one, no more resembles a Troupiale than it does a Thrush.

4. I shall exclude the Blue Bonana of Madras, because, on the one hand, it is inconsistent with the law of the climate, and on the other, the figure and description of Ray have nothing which would characterise the Bonana, not even the plumage. According to that author, its head, tail, and wings are blue, but the tail of a brighter tinge; the rest of the plumage black or cinereous, except the bill and the feet, which are rusty.

5. Lastly, I shall exclude the India Troupiale, not only on account of the difference of climate, but for other stronger reasons, which before induced me to place it between the Rollers and the Birds of Paradise.

Though we have ranged along with the Troupiales, the Cassies, the Baltimores, and the Bonanas, these, as they have received separate names, are distinguished by differences that are sufficiently important to form small subordinate *genera*. I am able, from the comparison of a number of these birds, to assert that the Cassies have the strongest bill, next to them the Troupiales, and then the Bonanas. With respect to the Baltimores, their bill is not only smaller than in the rest, but it is straighter, and of a peculiar shape. They seem also to have different instincts; I therefore retain their proper names, and treat of each separately.

The common characters which Briſſon aſcribes to them are the naked noſtrils, and the elongated conical form of the bill. I have already obſerved that the upper mandible extends over the cranium, or that the tuft, inſtead of making a point, makes a conſiderable re-entrant-angle; a circumſtance which ſometimes occurs in other ſpecies, but is moſt remarkable in the preſent.

The TROUPIALE*.

Oriolus Ictericus, Gmel.
Icterus, Briſſ.
Coracias Xanthornus, Scop.
The Yellow and Black Pye, Cateſby.
The Banana Bird from Jamaica, Albin.
The Icteric Oriole, Lath.

The moſt obvious features in the exterior appearance of this bird are its long pointed bill, the narrow feathers of its neck, and the great variety of its plumage: not only three colours enter into it, an orange-yellow, black and white, but theſe colours ſeem to multiply by their artful diſtribution. The black is ſpread over the head, the anterior part of the neck, the middle

* Briſſon gave this bird the Latin name *Icterus*, from its yellow colour; for the ſame reaſon Scopoli termed *Xanthornus*, or yellow bird; other authors have called it *Pica*, *Ciſſa*, *Picus*, and *Turdus*. The ſavages of Brazil name it *Guira-Tangeima*; thoſe of Guiana *Yapou*; the French coloniſts *Cul-jaune*, or yellow-bottom.

No. 72

THE ICTERIC ORIOLE.

middle of the back, the tail and the wings; the orange-yellow occupies the intervals, and all the under part of the body; it appears alſo in the iris *, and on the anterior part of the wings; the black which prevails through the reſt is interrupted by two oblong white ſpots, of which the one is placed at the coverts of the wings, the other on their middle quills.

The feet and nails are ſometimes black, ſometimes of a leaden colour. The bill ſeems to have no conſtant colour, for it has been obſerved to be in ſome white gray †, in others brown cinereous above ‡, and blue below; and laſtly, in others black above, and brown below §.

This bird is nine or ten inches long from the point of the bill to the end of the tail; and, according to Marcgrave, its wings extend fourteen inches, and its head is very ſmall. It is diſperſed through the region lying between Carolina and Brazil, and through the Caribbean iſlands. It is of the bulk of a blackbird; it hops like the magpie, and has many of its geſtures, according to Sloane. It has even, according to Marcgrave, the ſame cries; but Albin aſſerts that in all its actions it reſembles the Stare; and adds, that ſometimes four or five unite to attack a larger bird, which, after they have killed, they devour

* Albin adds, that the eye is encircled by a broad bar of blue; but he is the only perſon who has made this remark; it was probably an accidental variety.

† Briſſon. ‡ Albin. § Sloane and Marcgrave.

 orderly,

orderly, each maintaining his rank. Sloane, an author worthy of credit, ſays, that the Troupiales live on inſects. Yet there is no abſolute contradiction; for every animal which feeds on the ſmalleſt reptile is rapacious, and would feed on larger animals if it could do it with ſafety.

Theſe birds muſt be of a very ſocial diſpoſition since love, which divides ſo many other ſocieties, ſeems on the contrary to knit theirs more cloſely together. They do not ſeparate to accompliſh in retirement and ſecrecy the views of nature; a great many pairs are ſeen on the ſame tree, which is almoſt always lofty and ſequeſtered, conſtructing their neſt, laying their eggs, hatching and cheriſhing their infant brood.

Theſe neſts are of a cylindrical form, ſuſpended from the extremity of high branches, and waving freely in the air; ſo that the young are continually rocked. But ſome who believe that the birds act from deliberation, aſſert that the parents hang their neſt to avoid the attack of certain land animals, and eſpecially ſerpents.

The Troupiale is alſo reckoned very docile, and eaſily ſubjected to domeſtic ſlavery; which propenſity almoſt always attends a ſocial temper*.

* Specific character of the *Oriolus Ictericus*:—"Fulvous; the "head, throat, back, the wing-quills, and coverts, black, with a "white ſpot on the wings."

The ACOLCHI of SEBA*.

Oriolus Novæ Hispaniæ, Gmel.
Icterus Mexicanus, Briss.
The Mexican Oriole, Lath.

Seba, having found this name in Fernandez, has, according to his way, applied it arbitrarily to a bird entirely different from the one meant by that author, at least with regard to its plumage; and he has again ascribed to the same bird what Fernandez has said of the true *Acolchi*, which the Spaniards call *Tordo*, or Stare.

This false *Acolchi* of Seba has a long yellow bill; its head is all black, and also its throat; the tail and wings are blackish, and these are ornamented with small feathers of a golden colour, which have a fine effect on the dark ground.

Seba reckons his Acolchi an American bird, and I know not for what reason Brisson, who quotes no authority but Seba, subjoins that it is most common in Mexico. It is certain that the word *Acolchi* is Mexican, but we are not warranted to conclude the same thing of the bird on which Seba bestows it †.

* Contracted for *Acolchichi*.

† Specific character of the *Oriolus N. Hispania* :—" Yellow; " the head, upper part of the throat, wing-quills and tail, black; " the greater quills of the wings yellow at the tip, the lesser " all black."

The RING-TAILED ORIOLE, *Lath.*

L' Arc-en Queue, Buff.
Oriolus Annulatus, Gmel.
Icterus Caudâ Annulatâ, Briss.
Cornix Flava, Klein.

Fernandez gives the name of *Oziniſcan* to two birds which bear no reſemblance; and Seba has taken the freedom to apply the ſame name to a third entirely different from either, except in ſize, which is that of the pigeon.

The third *Oziniſcan* is the Ring-tail (*Arc-en-Queue*) of this article. I give it this name on account of a black arch or creſcent with its concavity turned towards the head, which appears diſtinctly on the tail when ſpread, and the more remarkable, as this is of a fine yellow colour, which is alſo that of the bill and of the whole body; the head and neck are black, and the wings of the ſame colour, with a ſlight tint of yellow.

Seba adds, that he received many of theſe birds from America, where they are looked upon as ravenous. Perhaps their habits are the ſame with thoſe of our Troupiales; the figure which Seba gives has a bill ſomewhat hooked near the point *.

* Specific character of the *Oriolus Annulatus*:—" Yellow; the " head and neck black, the greater coverts of the wings and their " quills edged with yellowiſh; the tail blackiſh and tinged."

The JAPACANI*.

Oriolus Japacani, Gmel.
Luſcinia pullo-lutea, Klein.

Sloane conſiders his *Little Yellow and Brown Fly-catcher* as the ſame with the Japacani of Marcgrave; but beſides the differences of the plumage, the Japacani is eight times larger, each dimenſion being double; for Sloane's bird is only four inches long, and ſeven over the wings, while Marcgrave's is of the bulk of the *Bemptere*, which is equal to that of the Stare, whoſe extreme length is ſeven inches, and its alar extent fourteen. It would be difficult to refer to the ſame ſpecies two birds, eſpecially two wild birds, ſo widely different.

The Japacani has a long pointed black bill, a little curved; its head is blackiſh, its iris of a gold colour; the hind part of its neck, its back, its wings, and its rump, are variegated with black and light brown; its tail is blackiſh below, and marked with white above; its breaſt, its belly, its legs are variegated with yellow and white, with blackiſh tranſverſe lines; its feet brown, its nails black and pointed †.

* This is the Brazilian name, according to Marcgrave.

† Marcgrave.

Sloane's little bird * has a round bill, almoſt ſtraight, and half an inch long; the head and back are of a light brown, with ſome black ſpots; the tail eighteen inches long, and of a brown colour, as alſo the wings, which have a little white at their tips. The orbits, the throat, the ſides of the neck, the coverts of the tail, yellow; the breaſt of the ſame colour, but with brown marks; the belly white; the legs brown, about fifteen lines long, and yellow in the toes.

This bird is common in St. Jago, once the capital of Jamaica; it lodges generally in the buſhes. Its ſtomach is very muſcular, and lined with a thin, looſe, inſenſible membrane. Sloane found nothing in the gizzard of the individual which he diſſected, but he obſerved that the inteſtines made a great number of circumvolutions.

The ſame author mentions a variety, which differs only becauſe it has leſs yellow in its plumage.

This bird may be reckoned a Troupiale, on account of the form of its bill; but it is certainly different from the Japacani †.

* *Oriolus Braſilianus*, Gmel.; *Icterus Braſilienſis*, Briſſ.; *Muſcicapa e fuſco et luteo varia*, Sloane; *the Brazilian Oriole*, Lath.

† Specific character of the *Japacani*:—"It is black and "duſky, below variegated with white and yellow, with tranſverſe "black lines, the head and tail blackiſh." Specific character of Sloane's little bird:—"It is yellow, the breaſt ſpotted, the head "and back marked by ſoft duſky ſpots, the belly white, the tail "and wings duſky, and the latter white at the tips."

The XOCHITOL and the COSTOTOL.

Oriolus Coſtototl, Gmel.
The New Spain Oriole, Lath.

Briſſon makes the *Xochitol* of Fernandez the tenth ſpecies of his *Troupiale of New Spain*, and which the Spaniſh naturaliſt conſiders as only the adult *Coſtotol*. But he takes notice of two Coſtotols, which are pretty much alike; but if they differ in ſome degree, we muſt refer what Fernandez ſays here to the Coſtotol of Chap. xxviii.

If we compare the deſcription of the Xochitol of Chap. cxxii. to that of Chap. xxviii. we ſhall meet with contradictions which it will be difficult to reconcile; for is it poſſible that the Coſtotol, which when ſo much grown as to be able to ſing is only of the ſize of a Canary bird, ſhould afterwards acquire the bulk of a Stare? that when young it has the ſweet warble of the Goldfinch, but after it is adult, and received the name of the Xochitol, it ſhould have the diſagreeable chatter of the Magpie? But wide differences alſo occur in the plumage; in the Coſtotol, the head and the under part of the body are yellow, while in the Xochitol they are black: in the former, the wings are yellow

tipped

tipped with black; in the latter, they are variegated with black and white above, and cinereous below, without a single yellow feather.

But all these contradictions will vanish, if, instead of the Xochitol of Chap. cxxii. we substitute the Xochitol or Flowery bird of Chap. cxxv. The size is nearly the same, being that of the Sparrow; its warble is pleasant, like that of the Costotol, the yellow of which is mingled with other colours that variegate the plumage of the former: they are both an agreeable food. The Xochitol resembles in two circumstances the Troupiales; it lives upon insects and seeds, and hangs its nest from the ends of small branches. The only difference which can be remarked between the Xochitol of Chap. cxxv. and the Costotol, is, that the latter is found in warm countries only, while the former inhabits all climates without distinction. But is it not likely that Xochitols go to breed in warm countries, where their young, or the Costotols, remain till they are grown up, or are Xochitols, and able to accompany their parents into colder climates? In the Costotol, the plumage is yellow, as I have said, and the tips of the wings black; and in the Xochitol of Chap. cxxv. the plumage is variegated with pale yellow, brown, white, and blackish.

Brisson has indeed made the latter his first *carouge*; but as it suspends its nest like the Troupiales, we have a decisive reason to range it it with theſe; except we reckon as another Troupiale the Xochitol of Chap. cxxii. of Fernandez, which is of the ſize of a Stare; its breaſt, belly, and tail of a ſaffron colour, variegated with a little black; its wings variegated with black and white below and cinereous above; its head, and the reſt of its body, black; it has the chatter of the Magpie, and its fleſh is good eating.

M

The TOCOLIN*

Oriolus Cinereus, Gmel.
Icterus Cinereus, Briſſ.
The Gray Oriole, Lath.

Fernandez conſidered this bird as a Wood-pecker, on account of its long and pointed bill; but this character belongs alſo to the Troupiales, nor can I perceive in the deſcription of Fernandez any other diſcriminating qualities of the Wood-peckers. I ſhall therefore leave it among the Troupiales, where Briſſon has placed it.

It is of the bulk of a Stare; it lives in the woods, and neſtles on trees. Its plumage is

* Its true name is *Ococolin*; but as I had appropriated it to another bird, I have here changed it, by prefixing the letter *T* of Troupiale.

beautifully

beautifully variegated with yellow and black, excepting the back, the belly, and the feet, which are cinereous.

The Tocolin is deſtitute of ſong, but its fleſh is good; it inhabits Mexico *.

* Specific character of the *Oriolus Cinereus*:—"It is variegated with yellow and black; its back, thighs, and belly, cinereous."

M

The COMMANDER.

Oriolus Phœniceus, Linn. and Gmel.
Icterus Pterophœniceus, Briff.
The Scarlet-feathered Indian Bird, Will.
The Red-winged Starling, Cateſby, Alb. and Kalm.
The Red-winged Oriole, Penn. and Lath.

This is the true Acolchi of Fernandez. It is called the Commander †, on account of a fine red mark on the anterior part of its wings, which in ſome meaſure reſembles the badge of the order of knighthood. The effect is here the more ſtriking, as it is thrown upon a ground of ſhining gloſſy black; for that is the general colour not only of its plumage, but of its bill, feet, and nails. There are, however, ſome ſlight exceptions; the iris is white, and the baſe of the bill is encircled by a narrow ring of red; the

† In Spaniſh, *Commendadoza*.

bill

bill alſo inclines ſometimes to brown rather than black, according to Albin. But the real colour of the mark on the wings is not a pure red, according to Fernandez, but is tarniſhed with a rufous tint, which increaſes and becomes at laſt the predominant colour. Theſe ſometimes ſeparate, the red occupying the anterior and the more elevated part of the ſpot; yellow, the poſterior and lower. But is this true with regard to all the individuals, or has not that been aſcribed to the whole ſpecies which is applicable only to the females? We are certain that in theſe the ſpot on the wings is not of ſo bright a red; beſides that diſtinction, the black of their plumage is mixed with gray, and they are ſmaller.

The Red-Wing is nearly of the ſize and ſhape of the Stare; its extreme length is eight or nine inches, and its alar extent thirteen or fourteen; it weighs three ounces and a half.

Theſe birds inhabit the cold as well as the warm countries: They are found in Virginia, Carolina, Louiſiana, Mexico, &c. They are peculiar to the New World, though one was killed in the environs of London; but this had doubtleſs eſcaped from its cage. They can be eaſily tamed, and taught to ſpeak; and they are fond of ſinging and playing, whether they be confined, or allowed to run through the houſe; for they are very familiar and lively.

The

The one killed near London was opened; in its ſtomach were found caterpillars, beetles, and maggots. But in America they feed on wheat, maize, &c. and are very deſtructive. They fly in numerous flocks, and, like the Stares of Europe, joining other birds equally deſtructive, as the Jamaica Magpies, they pour their famiſhed ſquadrons on the ſtanding crops and ſown fields; but the havock which they commit is by far greateſt in the warm regions, and near the ſea-coaſt.

When the planters fire on theſe combined flocks, birds fall of different kinds, and before the piece can be again charged, another flight arrives.

Cateſby informs us, that in Carolina and Virginia they always breed among the ruſhes; they interweave the points of theſe ſo as to form a ſort of roof or ſhed, under which they build their neſt, and at ſo proper a height that it can never be reached by the higheſt floods. This conſtruction is very different from that of our firſt Troupiale, and ſhews a different inſtinct, and therefore proves that it is a diſtinct ſpecies.

Fernandez pretends that they neſtle on the trees near the plantations: Has this ſpecies different cuſtoms accommodated to the different countries where it is found?

The Red-Wings appear in Louiſiana in the winter only, but they are then ſo numerous that three hundred have been ſometimes caught in a ſingle

ſingle draw of the net. For this purpoſe is uſed a long and very narrow net of ſilk, in two parts, like that for larks. " When they intend to " ſpread it," ſays Le Page Dupratz, " they " clear a place near the wood, and make a kind " of path, which is ſmooth and beaten, on which " they ſtrew a train of rice or other grain, and " retire to conceal themſelves behind a buſh " where the drag-cord is brought. When the " flocks of Red-Wings paſs over the ſpot, they " quickly deſcry the bait, light, and are caught " in an inſtant. It is neceſſary to diſpatch them, " it being impoſſible to collect ſo many alive."

But they are deſtroyed chiefly as being pernicious birds, for though they ſometimes grow very fat, their fleſh is always indifferent eating; another point of reſemblance to the Stares of Europe.

I have ſeen at Abbé Aubri's a variety of this ſpecies, in which the head and the upper part of the neck was of a light fulvous colour. The reſt of its plumage was the ſame as uſual. This firſt variety ſeems to ſhew that the bird repreſented in the *Planches Enluminées*, No. 343, by the name of *Cayenne Carouge*, is a ſecond, which differs from the firſt in wanting the red ſpots on the wings only; for the reſt of its plumage is exactly the ſame; the ſize is nearly alike, and the ſame proportions take place; and the difference between the climates is not ſo great, but

 that

that we may ſuppoſe a bird could be equally reconciled to both.

We need only compare No. 402, and Fig. 2, No. 236, of the *Planches Enluminées*, to be convinced that the bird engraved in the latter under the name of *Cayenne Troupiale*, is only a ſecond variety of the ſpecies of No. 402, under the name of *Red-winged Troupiales of Louiſiana*, which is the ſubject of the preſent article. It is nearly the ſame in ſize, ſhape, and in the kind and diſtribution of the colours; except that in No. 236, the red tinges not only the anterior part of the wings, but is ſpread over the throat, the origin of the neck, a portion of the belly, and even the iris.

If we next compare this bird, No. 236, with the one repreſented, No. 536, under the name of *Guiana Troupiale*, we ſhall perceive that the latter is a variety of the former, ariſing from the difference of age or ſex. All the colours are fainter; the red feathers are edged with white, and the black or blackiſh with light gray; ſo that the figure of each feather is diſtinctly marked, and the bird looks as if it were covered with ſcales. But the diſtribution of the colours is the ſame, the bulk the ſame, the climate the ſame, &c. It is impoſſible to diſcover ſo many relations ſubſiſting between birds of different ſpecies.

I am

I am informed that theſe frequent the Savannas, in the iſland of Cayenne, and commonly lodge in the buſhes, and that ſome people give them the name of *Cardinal**.

* Specific character of the *Oriolus Phœniceus*:—" It is black, " the coverts of the wings fulvous." In North America it is ſometimes called the *Swamp Black-Bird*. They appear in New-York about April, and retire in October to the South. In ſome of the colonies a premium of three-pence the dozen was offered for deſtroying them; and ſeveral farmers had the precaution to ſteep their Indian corn in a decoction of white hellebore before they ſowed it. Their neſt is ſuſpended among the buſhes and reeds in ſequeſtered ſwamps; it is ſtrong, covered externally with plaſtered broad graſs, and lined thickly with bent. The eggs are white, with ſcattered black ſtreaks.

M

The BLACK TROUPIALE.

Oriolus Niger, Gmel.
Icterus Niger, Briſſ.
Cornix parva profunde nigra, Klein.
The Black Oriole, Penn. and Lath.

The dark colour of this bird has procured it the names of Crow, Blackbird, and Daw.—But this is not ſo deep nor ſo uniform as has been alleged; the plumage in certain poſitions is of a black, changing with greeniſh reflexions, eſpecially on the head, the upper part of the body, the tail, and the wings.

It is of the ſize of a Blackbird, being ten inches long, and fifteen or ſixteen acroſs the wings, which when closed reach to the middle of the tail; this is four inches and a half in length, tapered, and consisting of twelve quills. The bill is more than an inch, and the mid-toe is longer than the leg, or rather the *tarsus*.

This bird is settled in St. Domingo, and is very common in some parts of Jamaica, particularly between Spanish-town and Passage fort. Its stomach is muscular, and generally contains caterpillars and other insects *.

* The specific character of the *Oriolus Niger* is, "That it is "entirely black." It inhabits also the whole extent of North America. They arrive at Hudson's Bay in June, and sing sweetly till their incubation, during which they only make a sort of chucking. They build their nests with grass and moss, on trees at the height of eight feet. Their eggs are five in number, dusky, and spotted with black. After hatching, they resume their warble; and collect in vast flocks to retire in September.

M

The Little BLACK TROUPIALE.

Oriolus Minor, Gmel.
The Lesser Black Oriole, Lath.

I have seen another Black Troupiale from America, but much smaller, and even inferior to the Red-wing Thrush in size: It was six or seven inches long, and its tail, which was square, only two inches and a half, and extended an inch beyond the wings.

The

The plumage was entirely black, but more gloſſy and floating with bluiſh reflexions on the head and the contiguous parts. It is ſaid that this bird can be eaſily tamed, and taught to live familiarly in the houſe.

The bird of No. 606, *Planches Enlumineés*, is probably the female of this; for it is entirely of a black or blackiſh colour, except the head and the tail, which are of a lighter tint, as is commonly the caſe in females. We alſo perceive the bluiſh reflexions which were remarked in the plumage of the male; but theſe appear not on the feathers of the head, but on thoſe of the tail and the wings.

No naturaliſt has, I apprehend, taken notice of this ſpecies *.

* Specific character of the *Oriolus Minor*:—" It is black, the " dark colour of the head mingled with cœrulean."

M

The BLACK-CAPPED TROUPIALE.

Le Troupiale a Calotte Noire, Buff.
Oriolus Mexicanus, Gmel.
The Black-crowned Oriole, Lath.

This bird appears to be entirely the ſame ſpecies with Briſſon's Brown Troupiale of New Spain. To form an idea of its plumage, imagine a bird of a fine yellow, with a black cap and mantle. The tail is of the ſame colour, and ſpotleſs; but the black on the wings is ſomewhat interrupted by the white which borders the coverts, and again appears on the tips of the quills. Its bill is of a light-gray, with an orange tinge, and the legs are cheſnut.—It is found in Mexico, and in the iſland of Cayenne *.

* Specific character of the *Oriolus Mexicanus*:—"It is blackiſh, "the upper ſide and the head yellow."

M

The SPOTTED TROUPIALE of Cayenne.

Oriolus Melancholicus, Linn. and Gmel.
Anthornus Nævius, Briſſ.
The Schomburger, Edw. and Lath.

The ſpots which occur in this ſmall Troupiale are owing to this circumſtance, that almoſt all the feathers, which are brown or blackiſh in the middle, are edged with yellow, more or leſs inclined to orange on the wings, the tail, and the lower part of the body. The throat is of a pure white; a ſtreak of the ſame colour which paſſes cloſe under the eye ſtretches back between two parallel black ſtreaks, one of which accompanies the white above, and the other bends round the eye below; the iris is of a bright a bright orange, almoſt red :—All theſe give a lively appearance to the male; for though the iris is orange alſo in the female, its plumage is of a tarniſhed yellow, which, mingling with a pale white, produces an unpleaſant uniformity.

The bill is thick and pointed, as in the Troupiales, and cinereous; the legs are fleſh-coloured. Its proportion may be conceived from the figure.

The ſpotted Carouge of Briſſon, which in many reſpects reſembles the Troupiale of this article, differs from it in ſeveral important circumſtances. It is not half the ſize, its hind nail is longer, its iris is hazel, its bill fleſh coloured, its throat, and the ſides of its neck, black; and laſtly, the belly, the legs above and below the tail are without a ſingle ſpot.

Edwards heſitated to which of two ſpecies he ſhould refer it; to the Thruſh, or to the Ortolan. Klein decides very readily, that it belongs to neither, but to the Chaffinch; yet notwithſtanding his deciſion, the ſhape of its bill, and the identity of the climate, determine me to adopt the opinion of Briſſon, who makes it a Carouge *

* Specific character of the *Oriolus Melancholicus*:—" It is " gray, dotted with black, with a white ſtripe on the eyes." Latham conſiders the Spotted Troupiale as a variety of this bird.

The OLIVE TROUPIALE of Cayenne.

Oriolus Olivaceus, Gmel.
The Cayenne Olive Oriole, Lath.

This bird is only ſix or ſeven inches long. It owes its name to the olive colour which prevails on the hind part of its neck, its back, its tail, its belly, and the coverts of its wings. But this colour is not uniform; it is darker on the neck, the back, and the adjacent coverts of the wings, and ſomewhat leſs ſo on the tail; it is much lighter under the tail, and alſo on a great part of the coverts of the wings, which are furtheſt from the back; with this difference between the large and the ſmall ſort, that the latter have no mixture of colour, while the former are variegated with brown. The head, the throat, the fore-part of the neck and the breaſt, are of a gloſſy brown, deeper under the throat, and inclining to orange on the breaſt, and running into the olive colour of the lower part of the body. The bill and legs are black; the wing-quills, and the large coverts neareſt the outer edge, are of the ſame colour, but bordered with white.

The

The ſhape of its bill is the ſame as that of the other Troupiales; its tail is long, and its wings when cloſed do not reach the third of the length *.

* Specific character of the *Oriolus Olivaceus*:—" It is olive, " the head, throat, and breaſt duſky, the wings black."

M

The WEAVER ORIOLE.

Le Cap-More †, Buff.
Oriolus Textor, Gmel.

The two birds figured No. 375 and 376, *Pl. Enl.* were brought by the captain of a ſhip who had collected forty birds from different countries, Senegal, Madagaſcar, &c. and who had called them Senegal Chaffinches. They have been termed Senegal Troupiales; but that appellation ſeems very improper; for the climate is different from that of the Troupiales, and the Weaver is widely diſtinguiſhed by the proportions of its bill, tail, and wings, and the manner in which it builds its neſt. It is perhaps the African repreſentative of the American ſpecies. The two which we have mentioned belonged to a lady of high rank, who al-

† The name *Cap-more* is contracted for *Capuchon-moraoré*, which denotes its *cowl of a duſky-golden gloſs*.

lowed them to be designed at her house, and has obligingly communicated some particulars that occurred with regard to the way in which they conducted themselves. This is the only source of information which we have.

The eldest had a kind of cowl which appeared of a brown-gold gloss in the sun; this cowl disappeared in the moulting during the autumn, leaving the head of a yellow colour; but it again returned in the spring, and was constantly renewed the succeeding years. The chief colour of the rest of the body was yellow, more or less inclined to orange; this was the predominant colour on the back, and on the lower part of the body; it bordered the coverts of the wings, their quills, and those of the tail, which were all of a blackish ground.

The young one had no cowl till the end of the second year, and did not even change its colours before that time; which occasioned its being mistaken for a female, and designed as such, No. 376. This mistake was excusable, since the distinction of sexes is not apparent during infancy, and one of the principal characters of the females is that of preserving long the marks of youth.

Before the change which took place in the colours of its plumage, the yellow was of a lighter tint than in the old one; it spread over the throat, the neck, the breast, and bordered, as in the other, all the quills of the tail and of the

the wings. The back was of an olive-brown, which extended beyond the neck as far as the head. In both the iris was orange, the bill of a horn colour, thicker and ſhorter than in the Troupiale, and the legs reddiſh.

Theſe two birds lived in the ſame cage, and at firſt upon good terms with each other; the young one ſat generally on the higheſt bar, holding its bill cloſe to the other, which it anſwered, by clapping its wings, and with a ſubmiſſive air.

They were obſerved in the ſpring to interweave chickweed in the grating of their cage; this was therefore conceived as an indication of their deſire to neſtle. They were ſupplied with ſmall ruſhes, and they built a neſt ſo capacious as to conceal one of them entirely. The following year they renewed their labour; but the young one being new clothed in the plumage of its ſex, was driven off by the other, and obliged to conduct its work alone in another corner of the cage. But it was ſtill perſecuted, and notwithſtanding its ſubmiſſive behaviour, it was often ſo roughly treated as to be left inſenſible. They were ſeparated, and each was intent on building; but the labours of one day were often deſtroyed in the ſucceeding—A neſt is not the production of an individual.

They had both a ſingular kind of ſong, ſomewhat ſhrill, but very ſprightly. The old one died ſuddenly, and the young one was cut off by by epileptic fits. Their ſize was rather inferior to that of our firſt Troupiale; and their wings and tail were alſo proportionably ſhorter *.

* Specific character of the *Oriolus Textor*:—" It is yellow; its " head duſky, gliſtening with gold; the quills of its wings and " tail blackiſh, and orange at their margin."

M

The WHISTLER.

Oriolus Viridis, Gmel.
Icterus minus Viridis, Briſſ.
The Whiſtler Oriole, Lath.

I ſee no reaſon why Briſſon has reckoned this bird a Baltimore, for both in the ſhape of its bill and in the proportions of its tarſus it ſeems more related to the Troupiales. But I leave the matter undecided, placing it between the Troupiales and Baltimores, and applying the vulgar name which it receives in St. Domingo, on account of its ſhrill notes.

This bird is in general brown above, except the rump and the ſmall coverts of the wings, which are of a greeniſh yellow, as alſo in the whole under-part of the body; but this colour is duſky below the throat, and variegated with ruſty on the neck and breaſt; the great coverts and the quills of the wings, as well as the twelve of the tail, are edged with yellow. But to form an an accurate idea of the plumage of the Whiſtler, we muſt imagine an olive tint of various intenſity ſpread over all the colours without exception. To characterize the predominant colour of the plumage of this bird, therefore, we ought to take olive and not green, as Briſſon has done.

The Whiſtler is of the ſize of a Chaffinch; it is about ſeven inches long, and ten or twelve inches acroſs the wings; the tail, which is unequally tapered, is three inches in length, and the bill nine or ten lines.

M

The BALTIMORE.

Oriolus Baltimore, Linn. and Gmel.
Icterus Minor, Briſſ.
Icterus ex auro nigroque varius, Klein.
The Baltimore Bird, Cateſby, Penn. and Lath.

This bird owes its name to ſome reſemblance that is perceived between the nature and diſtribution of the colours of its plumage, and the arms of Lord Baltimore *. It is a ſmall bird of the ſize of a houſe Sparrow, and weighing little more than an ounce; its length is ſix or ſeven inches, its alar extent eleven or twelve, its tail compoſed of twelve quills, and two or three

* Lord Baltimore was a Roman Catholic nobleman, who obtained the grant of Maryland, which he planted. T.

inches

inches long, ftretching more than a half beyond the wings when clofed. A fort of cowl of a fine black covers the head, and defcends before upon the throat, and behind as far as the fhoulders: the great coverts and the quills of the wings are alfo black, like thofe of the tail; but the former are edged with white, and the latter tipped with orange, which is the broader the farther they are from the mid-ones, in which it is wanting. The reft of the plumage is of a beautiful orange; and laftly, the bill and legs are of a lead colour.

In the female, which I examined in the Royal Cabinet, all the fore-part was of a fine black, as in the male, the tail of the fame colour, the great coverts and the wing-quills blackifh, the whole without any mixture of other colour; and what was fo beautiful an orange in the male, was only a dirty red in the female.

I have already faid, that the bill of the Baltimores was not only proportionably fhorter and ftraighter than in the Carouges, the Troupiales, and the Caffiques, but of a peculiar fhape: It is a pyramid of five fides, two belonging to the upper mandible, and three to the lower. I fhall add, that its leg, or rather its tarfus, is more flender than in the Carouges and Troupiales.

The Baltimores difappear in the winter, at leaft in Virginia and Maryland, where Catefby obferved them. They are alfo found in Canada, but Catefby met with none in Carolina.

They

They build their nefts on large trees, fuch as the poplars, the tulip trees *, &c. They fix it to the end of a thick branch, and commonly fupport it by two fmall fhoots which enter its fides; in which circumftance the nefts of the Baltimores feem to refemble thofe of the Loriots †.

* The tulip tree, *Liriodendron-Tulipifera*, LINN, is peculiar to America, and fo called, becaufe its flower-cup refembles a tulip in fize and fhape, and has fomewhat of the fame tints. T.

† Specific character of the *Oriolus Baltimore*:—" It is blackifh, " the under-part of its body, and a ftripe on its wings, fulvous." The neft is curioufly woven of tough filaments of plants, intermixed with wool, and lined with hair. It is pear-fhaped, open at top, with a hole in the fide, by which the young are fed and void their excrements. In fome parts of North America it is called, on account of its brilliancy, the *Fiery hang-neft*.

M

The BASTARD BALTIMORE.

Oriolus Spurius, Gmel.
Icterus Minor Spurius, Briff.
Tardus Minor gutture nigro, Klein.

This bird was no doubt fo called becaufe the colours of its plumage are not fo lively as in the Baltimore, and for this reafon it may be confidered as a degraded race. In fact, when we compare thefe birds, and find an exact correfpondence in every thing, except in the colours, and not even in the diftribution of thefe, but only in the different tints which they affume;

we

we cannot hesitate to infer that the Bastard Baltimore is a variety of a more generous race, degenerated by the influence of climate, or some other accidental cause. The black on the head is somewhat mottled, that of the throat pure; that part of the hood which falls behind is of an olive gray, which becomes darker as it approaches the back. Whatever in the preceding was bright orange, is in the present yellow, bordering on orange, and more vivid on the breast and the coverts of the tail than on any other place. The wings are brown, but their great coverts and their quills are of a dirty white. Of the twelve tail quills, the two central ones are blackish near their middle, olive at their origin, and yellow at their extremity; the next one on either side shews the two first colours mixed confusedly; and in the four following quills, the two last colours are melted together. In a word, the true Baltimore bears the same relation to the bastard one in respect to the colours of the plumage, that the latter bears to its female; in which the upper-part of the body is of a dusky white, and the under of a yellowish white. [A]

[A] Specific character of the *Oriolus Spurius*:—"It is black, "fulvous below, with a white stripe on the wings." In the State of New-York it usually arrives in May; attaches its nest to an apple-tree, and lays five eggs.

The YELLOW CASSIQUE of Brazil, or, the YAPOU.

Oriolus Persicus, Linn. and Gmel.
Cassicus Luteus, Briss.
The Black and Yellow Oriole, Lath.*

When we compare the Cassiques with the Troupiales, the Carouges, and the Baltimores, all which have many common properties, we perceive that they are larger, that their bill is stronger, and their legs proportionably shorter; not to mention the difference in the general appearance which it would be difficult to describe.

Several authors have given figures and descriptions of the Yellow Cassique under different names, and scarcely two of these exactly correspond.—But before we proceed to consider the varieties in detail, it will be proper to separate a bird, the characters of which seem to be widely distant from those of the Yellow Cassique of Brazil: It is the Persian Magpie of Aldrovandus. That naturalist describes it merely from a drawing, which had been sent from Venice. He reckons it to be of the size of our Magpie. Its predominant colour is not black,

* In Latin it has also been called *Pica, Picus Minor, Cissa, Nigra*, &c.; in Italian, *Gazza*, or, *Zalla di Terra Nuova*; in English, *The Black and White Daw of Brazil.*

but only duſkiſh (*ſubfuſcum*): Its bill is very thick, ſomewhat ſhort (*breviuſculum*) and whitiſh; its eyes white, and its nails ſmall; whereas the Yapou is ſcarcely larger than a Blackbird, and the dark part of its plumage is jet black; its bill is pretty long, of the colour of ſulphur; its iris is like ſapphire, and its nails of conſiderable ſtrength, according to Edwards, and even very ſtrong and hooked, according to Belon. We cannot doubt that birds ſo diſtinct belong to different ſpecies; eſpecially if Aldrovandus's information be true, that his bird is a native of Perſia, for we are certain that the Yapou is American.

The principal colours of the Yapou are conſtantly black and yellow, but the diſtribution is not uniformly the ſame, and varies in different individuals.—The one, for inſtance, which we have cauſed to be deſigned is entirely black, except the bill and the iris, as we have ſaid, and the great coverts of the wings neareſt the body, which are yellow, as alſo all the hind-part of the body, both above and below, from the thighs incluſively as far as the middle of the tail, and even beyond it.—In another, which was brought from Cayenne and lodged in the Royal Cabinet, and which is larger than the preceding, there is leſs yellow on the wings, and none at all on the lower part of the thigh, and the legs appear proportionably ſtronger:—it is probably a male.—In the Black and White Pye of Edwards, which

is

is evidently the ſame bird with ours, there is on four or five of the yellow coverts of the wings a black ſpot near their extremity; and beſides this, the black has purple reflexions, and the bird is rather larger.—In the Yapou or Jupujuba of Marcgrave, the tail is mottled with black and white only below, for its upper ſurface is entirely black, except the outmoſt feather on each ſide, which is yellow half its length.

It follows, therefore, that the colours of the plumage are by no means fixed and invariable in this ſpecies, which inclines me to believe with Marcgrave *, that the bird which Briſſon calls *the Red Caſſique*, is only a variety of the ſame.— I ſhall afterwards ſtate my reaſons †.

* I ſaw ſome entirely black, having the back of a blood colour. MARCGRAVE.

† Specific character of the *Oriolus Perſicus*:—" It is black, " the hind-part of its back, and a ſpot on the coverts of the wings, " and at the baſe of the coverts, yellow."

M

VARIETY *of the* YAPOU.

I.

The RED CASSIQUE of Brazil, or, the JUPUBA.

Oriolus Perſicus, var. 1. Gmel.

This is one of the names which Marcgrave gives to the Yapou, and which I apply to the Red

Red Caſſique of Briſſon, becauſe it reſembles that bird in the eſſential points; the ſame proportions, the ſame ſize, the ſame aſpect, the ſame bill, the ſame legs, and the ſame deep black diffuſed through moſt of its plumage. It is true, that the lower part of the back is red, inſtead of yellow, and the under ſurface of the body and of the tail entirely black; but this cannot be conſidered as a material diſtinction in a bird whoſe plumage, we have already obſerved, is ſubject to conſiderable variations. Beſides, yellow and red are contiguous colours, and apt to melt into orange; a circumſtance which may be occaſioned by difference of age, of ſex, of climate, or of ſeaſon.

Theſe birds are about twelve inches long, and ſeventeen acroſs the wings; the tail is forked and bluiſh; the two mandibles are equally arched downwards; the firſt *phalanx* of the outer toe in each foot ſeems to grow into the mid-toe; the tail conſiſts of twelve quills, and the under ſurface is white both below the black and the yellow part of the plumage.

They conſtruct their neſts with graſs, interwoven with horſe hair and hogs briſtles, or with vegetable productions which ſupply their place, and they imitate the form of a cucurbit fitted to its alembic. The neſts are brown on the outſide, and about eighteen inches deep, though the interior cavity is only a foot; the upper part is thick and prominent for the ſpace of half a foot; and here they are ſuſpended from the extremities

ties of ſmall branches. Sometimes four hundred of theſe neſts have been ſeen at once hanging in a ſingle tree, of the kind which the Brazilians call *Uti*; and as the Yapous hatch thrice a-year, the multiplication muſt be prodigious. This inſtinct of neſtling in ſociety on the ſame tree, marks ſome analogy to our Daws *.

* Linnæus and Gmelin conſider the bird deſcribed in this article as different from the *Jupujuba* of Marcgrave, and form it into a new ſpecies under the name of *Oriolus Hæmorrhous*, the *Red-rumped Oriole* of Latham. Its ſpecific character, "Black, with a "ſcarlet rump."

M

II.

The GREEN CASSIQUE of Cayenne.

Oriolus Criſtatus, var. 2. Gmel.

I ſhall not here be obliged to compare or diſcuſs the relations of other authors; for none has taken notice of this bird. Nor can I produce any information reſpecting its diſpoſitions and inſtincts. It is larger than the preceding; its bill is thicker at the baſe, and longer; and its legs, though ſtill as ſhort, would appear to be ſtronger. It has been very properly named the Green Caſſique, for all the fore-part both above and below, and even the coverts of the wings, are of that colour; the hind-part is cheſnut; the wing-quills are black, and thoſe of the tail partly

 black,

black, partly yellow; the legs are entirely black, and the bill is all red.

The length of this Caſſique is fourteen inches, and its alar extent eighteen or nineteen.

III.

The CRESTED CASSIQUE of Cayenne.

Oriolus Criſtatus, Gmel.
Anthornus Maximus, Pallas.
The Creſted Oriole, Lath.

This is alſo a new ſpecies, and the largeſt with which we are acquainted. Its bill is proportionably longer and firmer than in the others, but its wings are ſhorter. Its extreme length is eighteen inches, its tail five, and its bill two. It is alſo diſtinguiſhed from the preceding, by ſmall feathers, which it briſtles at pleaſure on the top of its head, and which form a ſort of moveable creſt. All the fore-part of this Caſſique, both above and below, including the wings and the legs, is black, and the whole of the reſt of a deep cheſnut. In the tail, which is tapered, the two middle quills are black, like thoſe of the wings, but all the lateral ones are yellow; and the bill is of the ſame colour.

I have ſeen in the Royal Cabinet, a ſpecimen which was rather of an inferior ſize, and in which the tail was entirely yellow; but I am not

not certain whether the two mid-quills were plucked, for it had only eight quills in all *.

* Specific character of the *Oriolus Cristatus*:—" It is very " black, its top somewhat crested by some elongated feathers."

IV.

The CASSIQUE of Louisiana.

Oriolus Ludovicianus, Gmel.
The White-headed Oriole, Penn. and Lath.

White, and changing violet, sometimes mixed together, sometimes separated, are all the colours of this bird. Its head is white, and also its tail, belly, and rump; the feathers of the wings and of the tail are of a waving violet, and edged with white; the rest of the plumage is dyed with a mixture of these colours.

It is a new species, lately brought from Louisiana. We may add, that it is the smallest of the Cassiques known; its whole length is only ten inches, and its wings when closed reach only to the middle of the tail, which is somewhat tapered.

The CAROUGE.

Oriolus Bonana, Linn. and Gmel.
Xanthornus, Briſſ. *
Turdus Minor Varius, Klein.
The Bonana Bird, Brown and Lath.

In general the Bonanas are ſmaller, and have a ſlenderer bill in proportion than the Troupiales. The ſubject of this article has its plumage painted with three colours, applied in large bodies.—Theſe are, 1. Reddiſh-brown, which is ſpread over all the fore-part of the bird, on the head, the neck, and the breaſt. 2. A velvet black on the back, the feathers of the tail, thoſe of the wings, and their great coverts, and even on the bill and the legs. 3. Deep orange on the ſmall coverts of the wings, the rump, and the coverts of the tail. All theſe colours are more obſcure in the female.

The length of the Bonana is ſeven inches, that of its bill ſix lines, that of its tail above three inches; its wings when ſpread meaſure eleven inches, and when cloſed extend to the middle of the tail, or beyond it. This bird was brought from Martinico; that of Cayenne

* Briſſon regards it as the ſame with the *Xochitol Altera* of Fernandez, already noticed. But its plumage is different, and though it inhabits the ſame country, it builds its neſt differently.

(Fig

(Fig. 1. No. 607, *Pl. Enl.*) is ſmaller, and the ſort of cowl which covers its head, neck, &c. is black, ſprinkled with ſome ſmall white ſpots on the ſides of the neck, and little reddiſh ſtreaks on the back; and laſtly, the great coverts and the middle feathers of the wings are edged with white. But theſe differences are, I conceive, too inconſiderable to prevent our ſuppoſing the Cayenne Bonana a variety of that of Martinico. They conſtruct a curious kind of neſt, reſembling the quarter of a hollow globe; and ſew it under the leaf of a Bonana, which ſhelters the neſt, and forms a part of it; the reſt conſiſts of the fibres of the leaves.

In what has been ſaid, it would be difficult to recogniſe the Spaniſh Nightingale of Sloane *; for that bird is in every reſpect ſmaller than the Bonana, being only ſix Engliſh inches in length, and nine acroſs the wings; its plumage is different, and it conſtructs its neſt in another mode. It is a ſort of bag, ſuſpended from the extremity of ſmall branches by a thread which they ſpin out of a ſubſtance that they extract from a paraſite plant, called *old man's beard*, which many have miſtaken for horſe-hair. In Sloane's bird the baſe of the bill was whitiſh, and encircled by a black ring; the crown of the head, the neck, the back, and the tail, were of a light

* Called alſo the *Watchy Picket* and *American Hang-neſt*. It is the *Oriolus Nidipendulus* of Gmelin, and the *Hang-neſt Oriole* of Latham.

brown, or rather reddiſh gray; the wings of a deeper brown, variegated with ſome white feathers, the lower part of the tail marked in its middle with a black line; the ſides of the neck, the breaſt, and the belly, of the colour of a dead leaf.

Sloane mentions a variety, either from age or ſex, which differs from the preceding, only becauſe its back has more of the yellow tint, the breaſt and belly of a brighter yellow, and there is a greater ſhare of black under the bill.

Theſe birds haunt the woods, and have an agreeable ſong. They feed on inſects and worms, for fragments of theſe are found in their gizzard or ſtomach, which is not muſcular. Their liver is divided into a great number of lobes, and of a blackiſh colour.

I have ſeen a variety of the St. Domingo Carouges, or the Yellow Bottoms of Cayenne, which I proceed to conſider: it reſembled much the female Bonana of Martinico, except that its head and neck were blacker. This confirms my idea, that moſt of theſe ſpecies are related, and that notwithſtanding our conſtant endeavour to reduce their number, we have ſtill carried the ſubdiviſions too far; eſpecially with regard to foreign birds, with which we are ſo imperfectly acquainted *.

* Specific character of the *Oriolus Bonana*:—" It is fulvous, " its head and breaſt cheſnut, its back and the quills of its wings " and tail, black."

The

The LESSER BONANA.

Le Petit Cul Jaune de Cayenne, Buff.
Oriolus Xanthornus, Linn. and Gmel.
Xanthornus Mexicanus, Briſſ. *

The male and female of this ſpecies are repreſented No. 5. fig. 1. and fig. 2. *Pl. Enl.* They have a jargon nearly like that of our Loriot, and ſhrill like that of our Magpie.

They ſuſpend their neſts, which are of a purſe ſhape, from the extremity of ſmall branches, like the Troupiales; but I am informed they chooſe the branches that are long and naked, and ſelect the trees that are ſtunted and ill-formed, and lean over the courſe of a river. It is alſo ſaid that theſe neſts are ſubdivided into compartments for the ſeparate families, which has not been obſerved in the Troupiales.

Theſe birds are exceedingly crafty, and difficult to enſnare. They are nearly of the ſize of a Lark; their length eight inches, their alar extent twelve or thirteen, the tail three or four inches, and tapered, ſtretching more than half beyond the cloſed wings. The principal colours of

* Briſſon ſuppoſes it to be the ſame with the *Ayoquantototl* of Fernandez, which has indeed the ſame ſize, and its plumage compoſed of black, yellow, and white. But Fernandez ſays nothing of the diſtribution of theſe colours, nor furniſhes any property characteriſtic of the ſpecies.

of thoſe repreſented No. 5, are yellow and black. In fig. 1. the black is ſpread over the throat, the bill, and the ſpace between that and the eye, the great coverts, and the quills of the wings, and of the tail, and the legs; all the reſt is yellow. But we muſt obſerve, that the middle quills and the great coverts of the wings are edged with white, and the latter ſometimes entirely white. In fig. 2. a part of the ſmall coverts of the wings, the thighs, and the belly, as far as the tail, are yellow, and the reſt all black *.

We may conſider, as varieties of this ſpecies, 1. The Yellow-headed American Carouge, or Bonana, of Briſſon. The crown of its head, the ſmall coverts of its tail, thoſe of the wings, and the lower part of the thigh, are yellow, the reſt of the body entirely black or blackiſh: it is about eight inches long, twelve inches acroſs the wings, the tail conſiſting of layers, containing twelve quills, each four inches long †. 2. The Bonana, or Carouge, of the iſland of St. Thomas, whoſe plumage is alſo black, except a little yellow ſpot on the ſmall coverts of the wings: it has twelve quills in the tail, which is tapered, as in the Leſſer Bonana, but ſomewhat longer. Edwards has deſigned one of the ſame ſpecies, Pl. 322,

* Specific character of the *Leſſer Bonana Bird*:—"It is yel-"low; the upper part of its throat, its tail, and wing-quills, "black."

† *Oriolus Chryſocephalus*, Linn. and Gmel. *Xanthornus Ictero-cephalus Americanus*, Briſſ. *The Golden-headed Oriole*, Lath.

which

which has a remarkable depreſſion at the baſe of the upper mandible *. 3. The Jamac of Marcgrave, which differs very little from it with reſpect to ſize, and of which the colours are the ſame, and diſtributed nearly in the ſame way as in fig. 1. except that the head is black, that the white on the wings is collected in a ſingle ſpot, and that a black line extends acroſs the back from the one wing to the other †.

* *Oriolus Cayanenſis*, Linn. and Gmel. *Xanthornus Cayanenſis*, Briſſ. *The Yellow-winged Pye*, Edw. *The Yellow-winged Oriole*, Lath.

† *Oriolus Jamacaii*, Gmel. *The Brazilian Oriole*, Lath.

M

The YELLOW-HEADED ORIOLE.

Les Coiffes Jaunes, Buff.
Oriolus Icterocephalus, Linn. and Gmel.
Xanthornus Icterocephalus Cayanenſis, Briſſ.
The Yellow-headed Starling, Edw.

Theſe are Cayenne Bonanas, which have a black plumage, and a ſort of cap that covers the head and part of the neck, but deſcends lower before than behind. A black ſtreak, which ſtretches from the noſtrils to the eyes, and turns round the bill, has been omitted in the figure. The ſubject repreſented Pl. 343, appears to be conſiderably larger than another which I have ſeen in the Royal Cabinet. Muſt this be aſcribed

to

to the difference of age, of ſex, of climate, or to the defect of the preparation? But from that variety Briſſon has drawn his deſcription: its ſize is equal to that of the Brambling: it is about ſeven inches long, and eleven acroſs the wings.

M

The OLIVE CAROUGE of Louiſiana.

Oriolus Capenſis, Gmel.
Xanthornus Capitis Bonæ Spei, Briſſ.
The Olive Oriole, Lath.

This bird is repreſented *Pl. Enl.* No. 607, Fig. 2, under the name of the Carouge (Bonana) of the Cape of Good Hope. I had long ſuſpected that this bird, though brought from the Cape to Europe, was really not a native of Africa; and the point is decided by the late arrival (October 1773) of a Bonana from Louiſiana, which is evidently of the ſame ſpecies, and differs in nothing but in the colour of the throat, which in the latter is black, and orange in the former. I am convinced that we ought to entertain the ſame opinion of all the pretended Bonanas and Troupiales of the ancient continent; and that we ſhall diſcover ſooner or later that they are either of a different ſpecies, or have derived their origin from America.

The

The Olive Bonana of Louiſiana has much of the olive tinge in its plumage, eſpecially on the upper part of the body; but this colour is not uniform; it is tinctured with gray on the crown of the head, and with brown behind the neck, on the back, the ſhoulders, the wings, and the tail; with a light-brown on the rump and the origin of the tail; and with yellow on the flanks and the thighs, and the large coverts and quills of the wings, whoſe fundamental colour is brown, are edged with yellow. All the under-part of the body is yellow, except the throat, which is orange; the bill and the legs are of a cinereous brown.

This bird is nearly of the ſize of a houſe ſparrow; its length ſix or ſeven inches, its alar extent ten or eleven inches. The bill is near an inch long, and the tail more than two; it is ſquare, and conſiſts of twelve quills. The firſt quill is the ſhorteſt of the wing, and the third and fourth the longeſt *.

* Specific character of the *Olive Oriole*:—"It is of a duſky "olive, below yellow."

The KINK.

Oriolus Sinensis, Gmel.
The Kink Oriole, Lath.

This new ſpecies, brought very lately from China, appears to reſemble ſo much the Bonana on the one hand, and the Blackbird on the other, that it may be regarded as the intermediate ſhade. The ſides of its bill are compreſſed as in the Blackbird, but not ſcalloped like thoſe of the Bonana; and Daubenton the younger has properly given it a diſtinct name, as being really different from theſe two ſpecies, though it connects the common chain.

The Kink is ſmaller than our Blackbird: its head, its neck, origin of its back, and its breaſt, are of aſh-gray, and this colour acquires a deeper hue as it approaches the back; the reſt of the body, both above and below, is white, as alſo the coverts of the wings, whoſe quills are of a poliſhed ſteel-colour, gliſtening with reflexions that play between greeniſh and violet. The tail is ſhort, tapered, and parted by this ſame ſteel colour and white; ſo that on the two mid-quills, the white is only a ſmall ſpot at their extremity; this white ſpot extends higher on the following quills, the farther they remove from the middle, and the ſteel colour retiring, is at laſt reduced on the two exterior quills to a ſmall ſpot near their origin.

N° 73

THE GOLDEN ORIOLE.

The LORIOT*.

Oriolus Galbula, Linn. and Gmel.
Oriolus, Briſſ.
Galbula, Ray, and Will.
Turdus Luteus, Friſch.
Turdus Aureus, Klein.
The Witwall, Will.
The Yellow-bird from Bengal, Alb.
The Golden Oriole, Penn. and Lath.

IT has been ſaid, that the young of this bird are excluded by degrees, and in detached parts, and that the firſt object of the parents is to collect and combine the ſcattered limbs, and, by virtue of a certain herb, to form them into an animated whole. The difficulty of this marvellous re-union hardly exceeds, perhaps, that of properly ſeparating the ancient names which the moderns have confuſedly applied to this ſpecies,

* In Greek, Χλωριον, from its greeniſh yellow colour; the female Χλωρις, according to Ælian; in modern Greek, Συκοφαγος, or fig-eater: In Latin it has alſo the names *Chlorion* and *Chloris*, beſides *Chloreus*, *Oriolus*; *Merula Aurea*, *Turdus Aureus*, *Luteus*, *Lutea*, *Luteolus*, *Ales Luridus*, *Picus nidum ſuſpendens*, *Avis Icterus*, *Galgalus*; and Pliny applies theſe four names, *Galbulus*, *Galbula*, *Vireo*, *Vineo*: in Italian, *Oriolo*, *Regalbulo*, *Gualbedro*, *Galbero*, *Reigalbero*, *Garbella*, *Rigeyo*, *Melziozallo*, *Becquafigo*, *Bruſola*: in Spaniſh, *Oropendula*, *Oroyendola*: in German, *Bierholdt*, *Bierolf*, *Brouder-Berolft*, *Byrolt*, *Tyrolt*, *Kirſcholdt*, *Gerolft*, *Kerſenriſe*, *Goldamſel*, *Goldmerle*, *Gut-merle*, *Olimerle*, *Gelbling*, *Widdewal*, *Witwal*: in Swiſs, *Wittewalch*. The name *Oriole* is derived from the Latin *Aureolus*, or Golden.

retaining thoſe which really belong to it, and referring the others to thoſe kinds which the ancients intended them to denote. I ſhall here obſerve only that, though this bird is diſperſed through a wide extent, there are certain countries which it ſeems to avoid. It is not found in Sweden, in England, in the Bugey mountains, nor in the heights of Nantua, though it appears in Switzerland regularly twice a year. Belon ſays that he never ſaw it in Greece; and how can we ſuppoſe that Ariſtotle knew this bird, without being acquainted with the ſingular conſtruction of its neſt, or if he knew it, that he ſhould have omitted to take notice of it?

Pliny ſpeaks of the *Chlorion* *, from the account of Ariſtotle; but is not always attentive to compare the information which he borrows from the Greeks, with what he draws from other ſources. He has mentioned the Loriot by four different terms †, without acquainting us whether it is the ſame bird with the *Chlorion*.

The

* Hiſt. Nat. lib. x. 29.

† " *Picorum* aliquis ſuſpendit in ſurculo (nidum) primis in ramis, cyathi modo." Lib. x. 33. " Jam publicum quidem omnium eſt *(galgulos)* tabulata ramorum ſuſtinendo nido providè eligere, cameràque ab imbri aut fronde protegere denſâ." – From this ſimilarity in the conſtruction of the neſt, we may conclude that the *Picus* and *Galgulus* are the ſame with the Loriot. That the *Galgulus* is ſtill the ſame with the *Avis Icterus* and the *Ales Luridus* appears from the two following paſſages: " *Avis icterus* vocatur a colore, quæ ſi ſpectetur, ſanari id malum (regium) tradunt, et avem mori; hanc puto Latine vocari *galgulum*." Lib. xxx. 11. " Icterias

The Loriot is a roving bird, continually changing its abode; it lives with us only during the ſeaſon of love. It obeys the primary impulſes with ardour and fidelity. The union is formed on the arrival, about the middle of the ſpring. The pair build their neſt on lofty trees, but often at no conſiderable height; they form it with ſingular induſtry, and in a way very different from that of the Blackbird, though they have been referred to the ſame genus. They commonly faſten to the fork of a ſmall branch long ſtraws or hemp-ſtalks; ſome of which, extending directly acroſs, form the margin of the neſt; others penetrate through its texture; while others, bending under it, give ſolidity to the ſtructure. The neſt is thus provided with an exterior cover, and the inner bed, prepared for receiving the eggs, is a matting of the ſmall ſtems of dog-graſs, the beards of which are ſo much concealed that the neſt has often been ſuppoſed to be lined with the roots of plants. The interſtices between the outer and inner caſe are filled with moſs, lichens, and other ſuch ſubſtances, which compact the whole. After the neſt is conſtructed, the female drops in it four or five eggs, the ground colour of which is a dirty white, and ſprinkled with ſmall diſ-

" Icterias (lapis) *aliti lurido* ſimilis, ideo exiſtimatur ſalubris contra regios morbos." *Lib.* xxxvii. 10. Beſides, in Book x. 25. Pliny ſays of the *Galgulus*, that " it retires as ſoon as it has reared its young," which agrees exactly with the Golden Oriole.

 tinct

tinct spots of a brown, approaching to black, most numerous on the small end. She sits closely three weeks, and not only retains long her affection * to her young, but defends them against their enemies, and even against man, with more intrepidity than could be expected from so small a bird. The parents have been seen to dart resolutely upon the plunderers of their brood; and what is still more remarkable, a mother, taken with her nest, continued to hatch in the cage, and expired on her eggs.

After the young are reared, the family prepares for its journey. This commonly happens in the end of August, or the beginning of September. They never assemble in numerous flocks, nor do the families remain united, for seldom are more than two or three found together. Though they fly rather heavily, flapping their wings like the Blackbird, they probably winter in Africa: for on the one hand, the Chevalier des Mazy, Commander of the Order of Malta, assures me, that they pass that island in the month of September, and repass it in the spring; and on the other, Thevenot says, that they migrate into Egypt in the month of May, and return in September †. He adds, that in May they are very fat, and their flesh good eating. Aldrovandus is surprised that in France they are never brought to our tables.

* Belon.

† Voyage du Levant, tom. i. p. 493.

The

The Loriot is about as large as the Blackbird; its length nine or ten inches, its alar extent ſixteen, its tail three and a half long, and its bill fourteen lines. The male is of a fine yellow over all the body, the neck, and head, except a black ſtreak which ſtretches from the eye to the corner of the aperture of the bill. The wings are black, except a few yellow ſpots, which terminate moſt of the great quills, and ſome of the coverts: the tail is divided by yellow and black, ſo that the black prevails on the part which appears of the two mid-quills, and the yellow gradually exends over the lateral quills, beginning at the tips of thoſe which are next the two middle ones. But the plumage is very different in the two ſexes. Almoſt all that was of a pure black in the male, is, in the female, of a brown, with a greeniſh tinge; and what was of a beautiful yellow in the former, is in the latter olive and pale brown:—olive on the head, and the upper part of the body dirty white, variegated with brown ſtreaks under the body, white at the tips of moſt of the wing-quills, and pale yellow at the extremity of their coverts; and there is no pure yellow, except at the end of the tail and on the lower coverts. I have beſides obſerved in a female, a ſmall ſpace behind the ear, without feathers, and of a light ſlate colour.

The young males reſemble the females with reſpect to plumage, and the more ſo the tenderer their age. At firſt they are ſtill more ſpeckled

 than

than the female, and even on the upper part of the body; but in the month of Auguſt the yellow begins to appear under the body. Their cry is different alſo from that of the old ones; they ſcream *yo*, *yo*, *yo*, ſucceeded ſometimes with a ſort of mewing like that of a cat *. But they have alſo a ſort of whiſtling, eſpecially before rain †; if this be not really the ſame with the mewing.

Their iris is red, the bill reddiſh brown, the inſide of the bill reddiſh, the edges of the lower mandible ſomewhat arched lengthwiſe, the tongue forked, and, as it were, jagged at the tip, the gizzard muſcular, terminating in a bag formed by the dilatation of the *œſophagus*, the gall bladder green, the *cæca* very ſmall and ſhort, and the firſt *phalanx* of the outer toe glued to that of the middle toe.

When they arrive in the ſpring, they feed on caterpillars, worms, inſects, whatever in ſhort they can catch; but they are fondeſt of cherries, figs ‡, the berries of the ſervice tree, peas, &c. A couple of theſe birds could in one day completely plunder a rich cherry-tree; for they peck

* Geſner ſays, that they pronounce *Oriot*, or *Loriot*; Belon, that they ſeem to ſay *compere loriot*, and others have fancied that they articulated *louſôt bonnes meriſes*, &c.

† Gesner.

‡ Hence they have been called Συκοφαγοι, and *Becafigos*. Perhaps the figs improve the quality of their fleſh; they do ſo in the caſe of Blackbirds.

the

the cherries one after another, and only eat the ripe part.

The Loriots are not eaſy to breed or tame. They can be caught by the call, placing limed twigs where they drink, and by various ſorts of nets.

Theſe birds have ſometimes ſpread from one end of the continent to another, without ſuffering any alteration in their external form, or in their plumage; for Loriots have been ſeen in Bengal, and even in China, which were preciſely like ours. But others have been brought from nearly the ſame countries, which had ſome differences in their colours, and which may be regarded, for the moſt part, as varieties of climate, till accurate obſervations, of their inſtincts, their habits, and manner of life, throw light on our conjectures *.

* Specific character of the *Golden Oriole*:—" It is yellow, its " ſtraps and joints are black, its outer tail-quills yellow behind." It ſeldom or never viſits England.

VARIETIES *of the* LORIOT.

I.

The COULAVAN.

Oriolus Chinensis, Linn. and Gmel.
Oriolus Cochinensis, Briss.

This bird is brought from Cochin-China: it is perhaps rather larger than our Loriot, its bill is also proportionably stronger; the colours of the plumage are precisely the same, and every where distributed in a similar manner, except on the coverts of the wings, which are entirely yellow, and on the head, where there is a sort of black horse-shoe, of which the convex part bounds the occiput, and its branches, passing below the eye, terminate in the corners of the opening of the bill. This is the most remarkable distinction of the Coulavan, and yet there is in the Loriot a black spot between the eye and the bill, which appears to be the rudiment of the horse-shoe.

I have seen some specimens of the Coulavan, in which the upper part of the body was of a brown yellow. In all, the bill is yellowish, and the legs black *.

* Specific character of the *Oriolus Chinensis*:—" It is yellow, " the joints black, but yellow at the tips, a black stripe on the " back of the head." Latham reckons it a variety.

II.

The CHINESE LORIOT.

Oriolus Melanocephalus, Gmel.
Sturnus Luteolus, Linn.
Oriolus Bengalensis, Briss.
The Black-headed Indian Icterus, Edw.

It is somewhat less than ours, but is of the same shape, proportions, and colours, though these are differently disposed. The head, the throat, and the fore-part of the neck, are entirely black *, and in the tail there is no black, but a broad stripe, which crosses the two intermediate quills near their extremity, and two spots placed very near the tips of the two following quills. Most of the coverts of the wings are yellow, the others are parted with black and yellow; the largest quills are black where they are seen when the wings are closed, and the others are edged or tipt with yellow; all the rest of the plumage is of the finest yellow.

The female is different †; for the front or the space between the eye and the bill is of a vivid yellow, the throat and the fore-part of the

* The sort of black piece that covers the throat and the fore-side of the neck, is in Edwards' figure a scallop on each near its middle.

† *Oriolus Galbula*, var. 1. Gmel. *Icterus Maderaspatanus Naevius*, Briss. *The Mottled Jay*, Ray. *The Yellow Starling from Bengal*, Alb. *The Yellow Indian Starling*, Edw.

neck of a light yellowiſh caſt, with brown ſpeckles; the reſt of the under-part of the body is of a deeper yellow, the upper of a ſhining yellow, all the wings variegated with brown and yellow, the tail alſo yellow, except the two mid-quills, which are brown, marked with a yellowiſh ſpangle, and tipt with yellow.

M

III.

The INDIAN LORIOT.

Oriolus Galbula, var. 2. Gmel.
Oriolus Indicus, Briſſ.
Chloris Indicus, Aldr.

It has more yellow than any of the Loriots, for it is entirely of that colour, except, 1. A horſe-ſhoe, which bends round the crown of the head, and terminates on each ſide in the corners of the bill. 2. Some longitudinal ſpots on the coverts of the wings. 3. A belt which croſſes the tail near the middle; the whole of an azure colour, but the bill and legs are of a glowing bright red.

M

IV.

The STRIPED-HEADED ORIOLE.

Le Loriot Rayé, Buff.
Oriolus Radiatus, Gmel.
Oriolus Capite ſtriato, Briſſ.
Merula Bicolor, Aldrov.

This bird has been regarded by ſome as a Blackbird, by others as a Loriot: its true place ſeems to be between the Loriots and the Blackbirds, and ſince its proportions are different from thoſe of either of theſe two ſpecies, I would conſider it as an intermediate or related ſpecies, rather than as a mere variety.

The radiated Loriot is not ſo large as a Blackbird, and of a more ſlender ſhape: its bill, tail, and legs, are ſhorter, but its toes longer; its head is brown, delicately radiated with white; its wing-quills are alſo brown, and edged with white; all the body is of a beautiful orange, deeper on the upper-part than on the lower; the bill and the nails are nearly of the ſame colour, and the legs are yellow.

The THRUSHES.

Les Grives, Buff.
Turdi, Linn. &c.

THE family of the Thrushes is certainly much related to that of the Blackbirds *; but it would be improper, as several naturalists have done, to confound them together. The common people appear to have acted more wisely, who have applied different names to objects which are really distinct. Those are termed Thrushes, whose plumage is speckled †, or marked with little strokes, disposed with a kind of regularity; on the contrary, those are Blackbirds whose plumage is uniform, or varied with large spots. We readily adopt this distinction, and reserving the Blackbirds for a separate article, we shall treat of the Thrushes in the present. We shall distinguish four principal species in our own climate, and to them we shall refer, as usual, their varieties and the foreign species most analogous.

The first species is the *Throstle*, Pl. Enl. No. 406; and I consider as varieties, the *White-headed*

* "Merulæ et turdi amicæ sunt aves," says Pliny. There seems little doubt that the Blackbirds and Thrushes consort, since they are commonly caught in the same snares.

† The word *grivelé* is used in the original, and is formed undoubtedly from *grive*, the term for a Thrush.

headed Thrush of Aldrovandus, the *Crested Thrush* of Schwenckfeld; and as foreign analogous species, the *Guiana Thrush*, Pl. Enl. No. 398, fig. 1. and *the Little American Thrush*, mentioned by Catesby.

The second species is the *Missel*, Pl. Enl. No. 489, which is the *turdus viscivorus* of the ancients, and to which I shall refer the *White Missel* as a variety.

The third species is the *Fieldfare*, Pl. Enl. No. 490; it is the *turdus pilaris* of the ancients. The varieties, the *Spotted Fieldfare* of Klein, and the *White-headed Fieldfare* of Brisson. I reckon as the analogous foreign kinds, the *Carolina Fieldfare* of Catesby, which Brisson makes his eighth species of Thrushes, and the *Canada Fieldfare* of Catesby, which Brisson makes his ninth species.

The fourth species is the *Red-Wing*, Pl. Enl. No. 51, which is the *turdus iliacus* of the ancients.

Lastly, I shall subjoin some foreign Thrushes, which are too little known to be referred to their proper species: such are the *Green Barbary Thrush* of Doctor Shaw, and the *Chinese Hoami* of Brisson, which I shall admit into the Thrushes, upon the authority of that naturalist, though it appears to me to differ from them in its plumage and in its shape.

Of the four principal species belonging to our climate, the two first, which are the Throstle and

and the Miſſel, reſemble each other. Both appear to be leſs ſubject to the neceſſity of migration, ſince they often breed in France, Germany, Italy, and in ſhort in thoſe countries where they paſs the winter. Both ſing delightfully, and they are of the ſmall number of birds whoſe warble is compoſed of a ſucceſſion of notes; and they both ſeem to be of an unſocial diſpoſition, for, according to ſome obſervers, they perform their journies alone. Friſch traces other analogies alſo between the colours of their plumage, and the order of their diſtribution, &c.

The two other ſpecies, *viz.* the Fieldfare and the Red-wing, are alſo analogous in ſome circumſtances. They travel in numerous flocks, are more tranſitory, and ſeldom neſtle in our climates; for which reaſon they ſing very ſeldom *, and their ſong is unknown not only to many naturaliſts, but even to moſt ſportſmen. It is rather a ſort of chirping, and when a ſcore meet on a poplar, they chatter all at once, making a very loud noiſe, which is far from being melodious.

Both ſexes of the Thruſh are nearly of the ſame ſize, and equally liable to change their plumage from one ſeaſon to another †. In all of

* FRISCH.—" In ſummer (ſays TURNER), the Turdus Pilaris is ſeldom or never ſeen with us (in England); in winter no birds are more numerous."

† " They have one colour in winter, another in ſummer." ARISTOTLE.

of them the firſt *phalanx* of the outer toe is joined to that of the mid-toe, the edges of the bill ſcalloped near the tip. None of them ſubſiſts on ſeeds; whether becauſe it ſuits not their appetite, or that their bill and ſtomach are too weak to break and digeſt them. Berries are their chief food, and hence they have received the epithet of *baccivorous*. They alſo eat inſects, worms, &c. and it is in queſt of theſe that they come abroad after rain, rove in the fields, and ſcrape the ground, eſpecially the Miſſels and the Fieldfares. They make the ſame ſearch in winter in places of a warm aſpect where the ground is thawed.

Their fleſh is a delicate food, eſpecially that of the firſt and fourth ſpecies, which are the Throſtle and the Red-Wing: but the ancient Romans held it in ſtill higher eſtimation than we, and kept theſe birds the whole year in a ſort of voleries, which deſerve to be deſcribed *.

Each volery contained many thouſand Thruſhes and Blackbirds, not to mention other birds excellent for eating, ſuch as Ortolans, Quails, &c. So numerous were theſe voleries in the vicinity of Rome, and in the territory of the Sabines, that the dung of the Thruſhes was employed to manure the lands, and what is remarkable, to fatten oxen and hogs †.

* "Inter aves turdus . . . Inter quadrupedes gloria prima lepus." MARTIAL.

† Varro, *De re Ruſticâ*. Lib. i. 31.

Theſe Thruſhes had leſs liberty in their voleries than our field pigeons in their dovecotes; for they were never ſuffered to go abroad, and they laid no eggs: but as they were ſupplied with abundance of choice food, they fattened to the great profit of the proprietor *. The voleries were a kind of vaulted courts, the inſide furniſhed with a number of rooſts. The door was very low, the windows were few, and placed in ſuch manner as to prevent the priſoners from ſeeing the fields, the woods, the birds fluttering at liberty, or whatever might awaken their ſenſibility, and diſturb the calm ſo conducive to corpulence. A little glimmering was ſufficient to direct them to their food; which conſiſted of millet, and a ſort of paſte made with bruiſed figs and flour. They had alſo given them the berries of the lentiſk, of the myrtle, of the ivy, and whatever in ſhort would improve the delicacy and flavour of their fleſh. They were ſupplied with a little ſtream of water, which ran in a gutter through the volery. Twenty days before they were intended for killing, their allowance was augmented; nay ſo far was the attention carried, that they gently removed into a little anti-chamber the Thruſhes

* Each fat Thruſh, except at the time of migration, ſold for three *denarii*, equal to about two ſhillings ſterling. And on the occaſion of a triumph or public feſtival, this ſort of trade yielded a profit of twelve hundred *per cent*. See Coſlumella, *de re Ruſticâ*, lib. viii. 10.—and Varro, lib. iii. 5.

which

which were plump and in good order, to enjoy more quiet; and frequently to heighten the illusion, they hung boughs and verdure imitating the natural scenery; so that the birds might fancy themselves in the midst of the woods. In short, they treated their slaves well, because they knew their interest. Such as were newly caught, were put in small separate voleries along with others that had been accustomed to confinement; and every contrivance, every soothing art, was employed to habituate them somewhat to bondage; yet these were birds never completely tamed.

We can at present perceive some traces of the ancient practice, improved indeed by the skill of the moderns. It is common in certain provinces of France to hang pots in the tops of trees which are haunted by the Thrushes; and these birds finding convenient sheltered nests, seldom fail to lay their eggs in them to hatch and rear their young *. This plan contributes doubly to the multiplication of the species; for it both preserves the brood, and by saving the time spent in building nests, it enables them to make two hatches in the year †. When they find no pots, they construct their

* Belon.

† It appears even that they sometimes have three hatches; for Salerne found in the month of September a Thrush's nest in a vine containing three eggs not yet hatched, which appeared to be of the third hatch.

nests

nefts in trees or even bufhes, and with great art; they cover the outfide with mofs, ftraw, dried leaves, &c. but they line the infide with a hard cafe formed of mud, compacted with ftraws and fmall roots. In this refpect they differ from the Pies and Blackbirds, which lay their eggs on a foft mattrefs. Thefe nefts are hollow hemifpheres about four inches in diameter. The colour of the eggs varies in the different fpecies between blue and green, with fome dull fpots that are moft frequent on the large end. Every fpecies has alfo its peculiar fong; and fometimes they have even been taught to fpeak *. But this muft be underftood chiefly of the Throftle and the Miffel, in which the organs of voice feem to be the moft perfect.

It is faid that the Thrufhes fwallow the berries entire of the juniper, the miffletoe, the ivy, &c. †, and void them fo little altered, that when they fall in a proper foil, they germinate and produce. But Aldrovandus affirms that, having made thefe birds fwallow the grapes of the wild vine and the berries of the miffletoe, he could never difcover in their excrements any of thefe that retained its form.

The Thrufhes have a ventricle more or lefs mufcular, no craw, nor even a dilatation of the

* "Agrippina, the wife of Claudius Cæfar, had a Thrufh which imitated human fpeech." Pliny, lib. x. 42.

† Linnæus.

æfophagus,

œsophagus which may ſupply its place, and ſcarce any *cæcum;* but all of them have a gall bladder, have the end of the tongue parted into two or three threads, and have eighteen quills in each wing, and twelve in the tail.

Theſe birds are ſad and melancholy, and as the natural conſequence of that diſpoſition, they are the more enamoured of liberty. They ſeldom play or even fight together; ſtill leſs will they bend to domeſtic ſlavery. But their love of freedom is not equalled by their reſources for preſervation. Their oblique and tortuous flight is almoſt their only protection againſt the ſhot of the ſportſman, or the talons of the bird of prey *. If they reach a cloſe branchy tree, they remain ſtill through fear, and can hardly be beat out †. Thouſands of them are caught in ſnares; but the Throſtle and the Red-Wing are the two ſpecies which can the moſt eaſily be caught by the nooſe, and almoſt the only ones that can be taken by the call.

Theſe nooſes are nothing but two or three horſe-hairs twiſted together, and forming a running knot. They are placed round the junipers or ſervice-trees in the neighbourhood of a fountain or a mere, and when the place is well choſen, and the ſprings properly ſet, ſeveral

* Skilful ſportſmen aſſure me that Thruſhes are difficult to ſhoot, even more ſo than Snipes.

† This is, perhaps, the reaſon that they are ſaid to be deaf; Κωφότερος κίχλης, *deafer than a Thruſh,* was a proverb in Greece. But all the fowlers aſſure me that the Thruſh has a very quick ear.

 hundred

hundred Thrushes have been caught in a day in the ſpace of a hundred acres.

It is aſcertained from obſervations made in different countries, that when the Thruſhes appear in Europe about the beginning of the autumn, they arrive from the countries of the north in company with thoſe numerous flocks of birds which, on the approach of winter, traverſe the Baltic ſea, and leave Lapland, Siberia, Livonia, Poland, and Pruſſia, for more temperate climates. So abundant are the Thruſhes then on the ſouthern ſhore of the Baltic, that, according to the computation of Klein, the ſingle city of Dantzic conſumes every year ninety thouſand pairs. It is equally certain that the ſurvivors which emigrate again after the rigors of winter, direct their courſe towards the north. But the different ſpecies arrive not all of them at the ſame time. In Burgundy, the Throſtle appears the firſt about the end of September, next the Red-Wing, and laſt of all, the Fieldfare and the Miſſel; but the latter ſpecies is much leſs numerous than the three others, which might be expected, ſince it is more diſperſed.

We muſt not ſuppoſe that all the ſpecies of Thruſhes paſs conſtantly in the ſame number; ſometimes they are very few, becauſe the ſeaſon has either been unfavourable to their multiplication, or to their migration*; at other times they are

* I am aſſured that ſome years the Red-Wings are very rare in Provence; and this is the caſe alſo in the northern countries.

are extremely numerous: and a very intelligent obſerver * has informed me, that he ſaw prodigious clouds of Thruſhes, chiefly Red-Wings and Fieldfares, alight in the month of March at Brie, and cover an extent of ſeven or eight leagues. This appearance, which was unexampled, laſted near a month, and it was remarked that the cold had continued very long that winter †.

The ancients ſaid that the Thruſhes came every year into Italy from beyond ſeas about the autumnal equinox, and that they returned about the vernal equinox, and that in both paſſages ‡ they aſſembled and reſted in the iſlets of Pontia, Palmaria, and Pandataria, which are nigh the Italian coaſts. They repoſe too in the iſland of Malta, where they arrive in October and November; the north-weſt wind brings ſome flocks, the ſouth or ſouth-weſt ſometimes beats them back. But they do not always arrive with certain winds, and their appearance depends oftener on the ſtate of the air than on its motion; for if, in calm weather, the ſky ſuddenly darkens with the preludes of a ſtorm, the ground is then covered with Thruſhes.

Nor does the iſland of Malta appear to limit the migration of the Thruſhes towards the

* Hebert.

† Letters of M. le Commandeur Godeheu de Riville, tom. i. pp. 91, 92. *Mem. Etran.*

‡ Varro, *De Re Ruſtica*, lib. iii. 5. Theſe iſlets lie ſouth of the city of Rome, ſomewhat to the eaſt: That of *Pandataria* is at preſent known by the name *Ventotene.*

ſouth; for they are found in the interior parts of the African continent, from whence they annually paſs, it is ſaid, into Spain *.

Thoſe which remain in Europe ſpend the ſummer in the mountain foreſts: and on the approach of winter, they remove from the heart of the woods where the fruits and inſects begin to fail, and ſettle on the ſkirts of the adjacent plains. It is, no doubt, during this flitting that in the beginning of November ſo great a number are caught in the foreſt of Compigne. It is uncommon, ſays Belon, to find the different ſpecies in numbers at the ſame time, and in the ſame place.

In all of them the edges of the upper mandible are ſcalloped near the point, the inſide of the bill is yellow, its baſe has ſome black hairs or briſtles projecting forwards, the firſt *phalanx* of the outer-toe is joined to that of the middle-toe, the upper-part of the body is of a deeper brown, and the under lighter and ſpeckled; laſtly, in all, or in moſt of them, the tail is

* "Being in Spain in 1707," ſays the tranſlator of Edwards, "in the kingdom of Valencia, on the ſea-coaſt, I ſaw in October great flocks of birds that came in a direct courſe from Africa. Some were killed, and found to be Thruſhes, but ſo dry and lean, that they had neither ſubſtance nor taſte. The people of the country told me that every year at the ſame ſeaſon ſuch flocks arrive, but that moſt of them proceed much further." Admitting the fact, I ſtill doubt whether theſe Thruſhes really come from Africa; for this would be contrary to their uſual route, and the tendency of their flight on their arrival is no proof of the direction of their whole courſe.

nearly

nearly a third of the total length of the bird, which varies in the different ſpecies between eight and eleven inches, and is only two-thirds of the alar extent; the wings when cloſed reach as far as the middle of the tail, and the weight of the bird is between two ounces and a half and four and a half.

Klein aſſerts, he is well informed that Thruſhes are found alſo in the northern parts of India, but which differ from ours in not migrating.

The THROSTLE*.

La Grive, Buff.
Turdus Musicus, Linn. and Gmel.
Turdus Minor, Briss.
Turdus in altissimis, Klein.
The Mavis, Throstle, or *Song-Thrush*, Will.

THIS species, in the French language, gives name to the whole genus. I have therefore ranged it in the first place, though in point of size it occupies only the third. It is very common in some parts of Burgundy, and called by the country people *The Little Thrush* †, or *Little Red-Wing* ‡. It commonly arrives every year about the time of vintage, probably attracted by the maturity of the grapes; and hence undoubtedly it has received the name of *Vine-Thrush*. It disappears during the frosts, and again makes a transient visit in the months of March or April before its migration in May. On the departure of the flock, they always leave a few stragglers behind, which are either unable to follow the main body, or, yielding to the

* In Greek, Κιχλα or Κυχλη: In Latin, *Turdus*: In Italian, *Tordo Mezzano*: In Spanish, *Zorzal*: In German, *Drossel*, or *Drostel* (hence the English name); and in Brandenburg, *Zippe*: In Poland, *Drozd*: In Smoland, *Klera*; and in Ostrogothia, *Klaedra*.

† *Grivette*.

‡ *Mauviette*.

the mild influence of ſpring, ſtop and breed in the foreſts that occur in their route *. This is the reaſon why ſome Throſtles conſtantly remain in our woods, where they build their neſt on the wild apple and pear-trees, and even in junipers and in the buſhes, as has been obſerved in Sileſia † and in England ‡. Sometimes they fix it in the trunk of a thick tree ten or twelve feet high, and prefer, for the materials, wood rotten and worm-eaten.

They generally pair about the end of winter, and form laſting unions. They make two hatches in the year, and ſometimes a third, when their former have not ſucceeded. The firſt laying conſiſts of five or ſix eggs, of a deep blue with black ſpots, moſt frequent at the large end; and in the ſubſequent hatches the number regularly diminiſhes. It is difficult in this ſpecies to ſeparate the males from the females; their ſize being the ſame in both ſexes, and the colours of their plumage, as I have ſaid, ſubject to vary. Aldrovandus ſaw, and cauſed to be delineated, three of theſe birds, caught in different

* Dr. Lottinger aſſures me that they arrive in the months of March and April in the mountains of Lorraine, and that they return in September and October. Hence it would follow that in theſe mountains, or rather in foreſts that cover them, they paſs the ſummer, and from theſe retreats viſit us in autumn. But muſt we apply this local remark of Lottinger's to the whole ſpecies? Obſervations alone will decide.

† Friſch. ‡ Britiſh Zoology.

 ſeaſons;

ſeaſons; all which differed in the colours of their bill, of their legs, and of their feathers: in one of them the ſtreaks on the breaſt were hardly perceptible. Friſch aſſerts, however, that the old males have a white ray above the eyes, and Linnæus makes theſe white eyelids one of the characters of the ſpecies. Almoſt all the other naturaliſts agree, that the young males can hardly be diſtinguiſhed but by their early inclination to chant: for the Throſtle ſings delightfully, eſpecially in the ſpring *, whoſe return it announces; and as it breeds ſeveral times in the year, it enjoys a ſucceſſion of the vernal pleaſures, and may be ſaid to warble three-fourths of the year. It ſits whole hours on the top of a tall tree, ſtraining its delicate throat. Its warbling conſiſts of ſeveral different couplets, like that of the Miſſel, but ſtill more varied and more charming; which has obtained for it in many countries the denomination of the *Singing Thruſh*. The ſong is undoubtedly intended to attract the female; for even the imperfect imitation of it will produce that effect.

Each brood follows ſeparately their parents; ſometimes ſeveral of theſe chancing to meet in the ſame wood, would induce us to think that

* On its firſt arrival, about the end of winter, it has only a feeble whiſtle, day and night, like the Ortolans.

they

they aſſociate in numerous flocks *; but their union is fortuitous and momentary; the families ſoon ſeparate, and even the individuals diſperſe after they are able to provide ſingly their ſubſiſtence †.

Theſe birds are found in Italy, France, Lorraine, England, Scotland, Sweden, where they haunt the foreſts which abound with maples ‡. They migrate from Sweden into Poland fifteen days before and after the feſtival of St. Michael, when the weather is warm and calm.

Though the Throſtle is quick ſighted, and very alert to avoid its declared enemies, and to eſcape from manifeſt dangers; it has at bottom but little cunning, and is quite unguarded againſt concealed ſtratagems: it is eaſily caught either by the call or the gin, though leſs ſo than the Red-Wing. In ſome parts of Poland, ſuch numbers are taken that ſmall barks are loaded with them for exportation §. It is a bird that delights in woods, and in ſuch places the ſnares may be laid with ſucceſs. It ſeldom is met with in the plains, and even when it viſits the vines it conſtantly retires into the neighbouring copſes in the evening, and during the heat of the day;

* Friſch.—Dr. Lottinger alſo ſays, that though they do not migrate in troops, many are found together or pretty nigh each other.

† I am aſſured, however, that they like the company of the Calendar Larks.

‡ Linnæus, *Fauna Suecica.*

§ Rzaczinſki.

ſo

ſo that to ſucceed in catching the Throſtle, we ought to chooſe the proper time; its departure in the morning, and its return in the evening, or the mid-day, when the ſun's rays are moſt oppreſſive. Sometimes they are intoxicated with eating ripe grapes, and then they fall an eaſy prey.

Willughby informs us, that this ſpecies breeds in England, and reſides there the whole year; and he adds, that its fleſh is excellent, but partakes of the quality of its food. Our Throſtle ſubſiſts in autumn on cheſnuts, beech-maſt, grapes, figs, ivy-berries, juniper-berries, the fruit of the ſervice-tree, and ſuch like aliments. We are not ſo certain what it lives upon in the ſpring. In that ſeaſon it commonly appears on the ground in the woods, in wet places, and among the buſhes which ſkirt the flooded meadows, where it may be ſuppoſed to ſearch for earth-worms, ſnails, &c. If an intenſe vernal froſt happens, the Throſtles, inſtead of flying to milder climates, retire to the ſprings, and languiſh and pine; and a continuance of this ſevere weather will deſtroy many of them. This would ſeem to ſhew, that cold is not the ſole cauſe of their migrations, but that they have a certain circuit to deſcribe annually in a given time. It is ſaid that pomegranates prove a poiſon to them. In Bugey, the neſts of the Throſtles are much ſought after, or rather their

their young, which are dreſſed into delicate diſhes.

I ſhould ſuppoſe that this ſpecies was unknown to the ancients; for Ariſtotle reckons only three kinds *, which are all different from the preſent, and of which we ſhall treat in the following articles. Nor can we imagine that Pliny meant this when he ſpeaks of a new ſpecies which appeared in Italy in the time of the war between Otho and Vitellius; for that bird was almoſt as large as a Pigeon †, and therefore four times the ſize of the Throſtle, which weighs only three ounces.

I have obſerved in a Throſtle which lived ſome time with me, that when it was angry it cracked and ſnapped with its bill; its upper mandible was alſo moveable, though much leſs than the lower; alſo its tail was ſomewhat forked, which is not very evident from the figure ‡.

* *Hiſt. Anim.* lib. ix. 20.

† Pliny, lib. x. 49.

‡ Specific character of the Throſtle, *Turdus Muſicus*, LINN.—" Its wing-quills are ferruginous at their inner baſe." It is nine inches long, and its alar extent thirteen and a half. It ſings, eſpecially in the evenings of the ſpring, from the top of the higheſt tree; and breeds in buſhes and thickets. Its neſt is formed with earth, moſs, and ſtraws, the inſide plaſtered with clay. It lays five or ſix eggs of a bluiſh green, variegated with a few black ſpots.

VARIETIES *of the* THROSTLE.

I. The WHITE THROSTLE. The ſole difference conſiſts in the whiteneſs of its plumage; a quality which, though commonly aſcribed to the influence of the northern climates, may be produced by accidental cauſes in the more temperate countries, as we have remarked in the hiſtory of the Raven: but this colour is not ſpread over the whole body, nor is it pure. The breaſt and neck are marked with the ſpeckles peculiar to the Thruſhes, though far more dilute and faint; the back is ſhaded with a mixture of brown, and the breaſt tinged with rufous, as in thoſe figured by Friſch, pl. 33, but without any deſcription. Sometimes none of the upper part of the body, except the head, is white, as in that deſcribed by Aldrovandus; at other times the lower part of the neck only is marked by a white-croſs-bar, like a half collar; and, in different individuals, this colour certainly mingles variouſly with thoſe proper to the ſpecies.—But diſtinctions of that ſort cannot form even permanent varieties.

II. The CRESTED THRUSH, mentioned by Schwenckfeld, muſt alſo be regarded as a variety of this ſpecies; not only becauſe it is of the ſame ſize, and its plumage ſimilar, except

cept a whitiſh tuft, formed like that of the Creſted Lark, and alſo its collar white, but becauſe it is very rare. It may be even ſaid, that hitherto it is *unique*, ſince Schwenckfeld is the only perſon who has ſeen it, and that only once when it was caught in 1599 in the foreſts of the Dutchy of Lignitz.—It may be proper to mention, that theſe birds have ſometimes a creſt formed in drying, from the contraction of certain muſcles of the ſkin which cover the head.

FOREIGN BIRDS,

THAT ARE RELATED TO THE THROSTLE.

I.

The GUIANA THRUSH.

La Grive de la Guyane, Buff.
Turdus Guianensis, Gmel.

THE coloured figure conveys nearly all the information vhich we possess with regard to this little bird. Its tail is longer, and its wings proportionably shorter than in the Throstle; but its colours are nearly the same, only the speckles are spread as far as the last of the inferior coverts of the tail.

As the Throstle visits the countries of the north, and is besides very fond of changing its residence, it may have thence migrated into North America, and penetrated towards the south, where it would experience the alterations produced by the difference of climate and of food *.

* Specific character of the *Turdus Guianensis*:—"Above it is "dusky-greenish, below okery-white, with black longitudinal "streaks."

II. The

II.

The LITTLE THRUSH.

La Grivette d'Amerique, Buff.
Turdus Minor, Gmel.
Turdus Iliacus Carolinensis, Briss.
Turdus Minimus, Klein.
Merula Fusca, Sloane.

This bird occurs not only in Canada, but in Pennsylvania, Carolina, and as far as Jamaica: it spends only the summer in the northern provinces; though in the milder regions of the south it resides the whole year. In Carolina it haunts the thickest woods contiguous to the swamps; but in the hotter climate of Jamaica it retires to the forests that cover the mountains.

The specimens described or figured by naturalists differ in the colours of their feathers, of their bill, and of their legs; which would imply (if they all belong to the same species), that the plumage of the American Throstles is no less variable than those of Europe, and that they all spring from a common stem. This conjecture derives force from the numerous analogies which this bird has to the Thrushes, in its shape, in its port, in its propensity to migrate, and to feed upon berries, in the yellow colour of its internal parts, observed by Sloane, and in the speckles

 which

which appear on its breaſt; but it ſeems the moſt nearly related to our Throſtle and Red-Wing, and a compariſon of the points of ſimilarity is neceſſary to determine the ſpecies to which it belongs.

This bird is ſmaller than any of our Thruſhes, as in general are all the birds of America, if compared with their archetypes in the old continent. Like the Red-Wing, it does not ſing, and has fewer ſpeckles than that ſpecies, and therefore than any of the genus; like the Red-Wing alſo, its fleſh is delicate.—So far the American Thruſh reſembles the Red-Wing, but it has more numerous relations to our Throſtle; and, in my opinion, more deciſive ones. It has beards round the bill, a ſort of yellowiſh plate on the breaſt; it readily ſettles and remains in a country which affords it ſubſiſtence; its cry is like the winter-notes of the Throſtle, and therefore unpleaſant, as generally are the cries of all birds that live in wild countries inhabited by ſavages. Beſides, the Throſtle, and not the Red-Wing, is found in Sweden, whence it could eaſily migrate into America.

This Throſtle arrives in Pennſylvania in the month of May; it continues there the whole of the ſummer, during which time it hatches and raiſes its young. Cateſby tells us, that few of theſe Throſtles are ſeen in Carolina, whether becauſe a part only ſettle of what arrive, or that, as we have already obſerved, they conceal them-

ſelves

No. 74

THE REED THRUSH.

ſelves in the woods. They ſubſiſt on the berries of the holly, of the white-thorn, &c.

In the ſpecimens deſcribed by Sloane, the noſtrils were wider, and the feet longer than in thoſe deſcribed by Cateſby and Briſſon. Nor was their plumage the ſame; and if theſe differences were conſtant, we ſhould have reaſon to conclude that they belong to another family, or at leaſt are a permanent variety of this ſpecies. [A]

[A] Gmelin and Latham make the American and the Jamaica Throſtle to be two different ſpecies. The former, *Turdus Minor*, is thus characterized by Latham:—" It is duſky-rufous, the under-ſide of its body variegated with blackiſh ſpots; the upper-part of its throat, its belly, and its vent, white." The ſpecific character of the latter, the *Turdus Jamaicenſis*, or *Jamaica Thruſh*: "—It is duſky-cinereous, below white, its throat ſtriated longitudinally with brown, its breaſt cinereous."

M

III.

The REED THRUSH*.

La Rouſſerole, Buff.
Turdus Arundinaceus, Linn. Gmel. and Briſſ.
Junco, Geſner, Aldrov. Ray, and Will.

This bird has been called the River Nightingale, becauſe the male chants night and day,

* In Latin it has been called *Junco*, *Cinclus*, *Turdus Paluſtris* (Marſh-Thruſh), *Paſſer Aquaticus* (Water-Sparrow): in Italian, *Paſſere*

day, while the female is employed in hatching, and becaufe it haunts wet places. But though its fong has a greater extent, it is far from being fo pleafant as that of the Nightingale. It is commonly accompanied with a very brifk motion, and a trembling of the whole body. The bird climbs like the Creepers along the reeds and the low willows in fearch of infects, which conftitute its food.

The habit of this bird in frequenting the marfhes would feem to exclude it from the family of the Thrufhes; but it refembles them fo much in its external form, that Klein, who faw one almoft alive, fince it was killed in his prefence, doubts whether it could be referred to another genus. He informs us, that thefe birds inhabit the iflands in the mouth of the Viftula, and make their neft on the ground along the fides of the little hillocks covered with mofs *. He fufpects that they pafs the winter in the denfe marfhy forefts †; and he adds, that the upper-part of their body is a rufous brown, the lower of a dirty white, with fome afh fpots; the

Paffere d'Acqua: in the language of the American Indians, *Atototloquichitl*, according to Nieremberg; *Acototloquichitl*, according to Fernandez; *Caracura*, according to Laet The French name *Rouſſerole* is derived from *rouſſe*, the feminine of *roux*, which denotes its rufous plumage.

* It builds among the canes, fays Belon, with fmall reed ftraw, and lays five or fix eggs.

† Belon at firft fuppofed the Reed-Thrufh to be a bird of paffage, but afterwards difcovered the contrary.

bill black, the inside of the mouth orange, as in the Thrushes, and the legs lead-coloured.

An intelligent observer has assured me that he was acquainted in Brie with a small bird of this kind, and vulgarly called *Effarvatte*, which also prattles continually, and lodges among the reeds like the other. This reconciles the opposite opinions of Klein and Brisson with regard to the size of the Reed-Thrush; the former maintaining that it is as large as a Throstle, the latter that it does not exceed the Lark. It flies heavily, and flaps with its wings; the feathers on its head are longer than the rest, and form an indistinct crest.

Sonnerat brought from the Philippines a true Reed-Thrush, exactly similar to that of No. 513 *.

* Specific character of the *Turdus Arundinaceus* :—" It is dusky-" ferruginous, below of a chalky-white, its wing-quills marked " with tawny stripes at the tips." It is larger than a Lark, being seven inches long. Its eggs are yellowish-white, with dusky spots.

M

IV.

The MISSEL THRUSH*.

La Draine, Buff.
Turdus Viscivorus, Linn. Gmel. Gesner, Aldrov. &c.
Turdus Major, Briss.
Turdus Viscivorus Major, Ray.
The Miseltoe-Thrush, or *Shreitch*, Charl.
The Missel-Bird, or *Shrite*, Will.

The Missel weighs five ounces, and is distinguished by its magnitude from all the other Thrushes: but it is far from being so large as a Magpie, which Aristotle is made to assert †; an error probably of the copyist; or perhaps it attains to a greater size in Greece than with us.

The Greeks and Romans considered the Thrushes as birds of passage ‡, not excepting the Missel, with which they were perfectly acquainted under the name of *viscivorous* Thrush, or *feeder on misletoe-berries* §.

In

* In Greek, Ιξοσορος, or Μυρτοπαλλος: in Turkish, *Garatauk*: in Italian, *Tordo*, *Turdela*, *Gardenna*, *Dressa*, *Dressano*, *Gasotto*, *Columbina*: in German, *Zeher*, *Zerner*, *Ziering*, *Schnarre*, *Schnerrer*: in Polish, *Orozd*, *Naywiekssky*, *Jemiolucha*, *Cnapio*: in Welch it is called *Pen y llwyn*, which signifies *master of the coppice*.

† *Hist. Anim.* lib. ix. 20.

‡ Aristot. *Hist. Anim.* lib. viii. 16.—Pliny, lib. x. 24.—Varro *De Re Rusticâ*, lib. iii. 5.

§ Bird-lime (*viscus*) was formerly made from the berries of the misletoe: hence the Latin proverb *Turdus malum sibi cacat*; that parasite

Nº 75

FIG. 1. THE MISSEL BIRD FIG. 2. THE THROSTLE.

In Burgundy, the Miſſels arrive in flocks about the months of September and October, coming moſt probably from the mountains of Lorraine *. Part of them purſue their journey, and depart always in numerous bodies in the beginning of winter, while the reſt remain till the month of March; for ſome of them always continue during the ſummer both in Burgundy, and in other provinces of France, of Germany, of Poland †, &c. In Italy alſo, and in England, ſo many neſtle that Aldrovandus ſaw the new brood ſold in the markets; and Albin regards the Miſſels as not birds of paſſage ‡. Thoſe

paraſite plant being ſuppoſed to be propagated on the oak from the berries voided by the Miſſels. Bird-lime is now obtained by macerating the inner bark of holly.

* Dr. Lottinger of Sarbourg aſſures me, that ſuch of the Thruſhes as remove from the mountains of Lorraine on the approach of winter, depart in September and October, and return in the months of March and April; and that they breed in the foreſts with which the mountains are covered, &c. This account agrees well with what we have given from our own obſervation. But I muſt confeſs that another remark of that gentleman diſagrees with one of a very intelligent ornithologiſt: The latter (M. Hebert) pretends that in Brie the Thruſhes aſſemble at no time of the year; whereas M. Lottinger aſſerts that in Lorraine they always fly in flocks, and in fact they ſeem to arrive in troops about Montbard, as I have remarked. Can their habits differ in different countries, or in different times? This is not altogether unexampled; and I muſt add, from a more particular obſervation, that after the November paſſage is over, thoſe which remain in our diſtricts live ſeparately till the time of hatching: in ſhort, that the aſſertions of theſe two obſervers may be true, with certain reſtrictions.

† Rzaczynſki.

‡ The authors of the Britiſh Zoology are of the ſame opinion.

 which

which remain lay and hatch ſucceſsfully. They build their neſts, ſometimes in trees of a middling height, and ſometimes on the top of ſuch as are extremely tall, but always prefer thoſe which are moſt covered with moſs. They conſtruct both the inſide and outſide with herbage, leaves, and moſs, eſpecially the white moſs; and their neſt reſembles more that of the Blackbird than of the other Thruſhes, except its being lined with bedding. They lay four or five gray-ſpotted eggs; they feed their young with caterpillars, worms, ſlugs, and even ſnails, the ſhell of which they break. The parents eat all ſorts of berries during the ſummer, cherries, grapes, olives, the fruits of the cornel and the ſervice-trees; and in winter they ſubſiſt upon the berries of the juniper, of the holly, of the ivy, of the buck-thorn; upon beech-maſt, ſloes, fennel, and, above all, upon miſletoe berries. When diſturbed they cry *tré*, *tré*, *tré*; hence their name in the dialect of Burgundy *draine* and even ſome of the Engliſh names. In the ſpring the females have no other notes; but the males, ſitting on the tops of the trees, ſing charmingly, and their warble conſiſts of different airs that form a conſtantly varied ſucceſſion. In winter they are no longer heard. The male differs not in external appearance from the female, except that he has more black in his plumage,

Theſe

Theſe birds are of a gentle pacific temper; they never fight with one another, but yet are anxious for their own ſafety. They are more cautious even than the Blackbirds, which are generally reckoned very ſhy and timorous; for theſe are ſometimes decoyed by the call, while the Miſſels reſiſt the allurement. They are, however, caught ſometimes in gins, though leſs frequently than the Throſtles or Red-Wings.

Belon aſſerts, that the fleſh of the Miſſel, which he calls the Great Thruſh, is of a ſuperior flavour to that of the other ſpecies; but this is contrary to the account of all other naturaliſts, and to my own experience. Our Miſſels live not indeed upon olives, nor our ſmall Throſtles upon miſletoe-berries, as thoſe of which he ſpeaks; and it is well known how much the difference of food affects the quality of game *.

* Specific character of the *Turdus Viſcivorus*:——"Its back "is duſky, its neck ſpotted with white, its bill yellowiſh." The Miſſel Thruſh reſembles much the Throſtle, but the ſpots on its breaſt are large and round, not elongated as in the latter; and the inner coverts of its wings are white, thoſe of the Throſtle yellow. It is alſo of ſuperior ſize; for it is eleven inches long, ſixteen and a half broad, and weighs near five ounces. It builds its neſt in buſhes, or in the ſides of trees, particularly the aſh.

VARIETY *of the* MISSEL THRUSH.

The only variety I find in this ſpecies is the Whitiſh Miſſel noticed by Aldrovandus. The quills of its tail and wings were of a light and almoſt whitiſh colour, the head and all the upper-part of the body cinereous.

We may remark in this variety the alteration of the colour of the quills, of the wings, and of the tail, which are commonly ſuppoſed to be the leaſt liable to change, and as being of a deeper dye than the other feathers.

I may add that there are always ſome Miſſels which breed in the Royal Garden on the leafleſs trees; they ſeem to be very fond of yew berries, and eat ſo plentifully of them that their excrements are red; they are alſo attached to the fruit of the lote.

In Provence the people have a ſort of call with which they imitate the vernal ſong of the Miſſel Thruſh and of the Throſtle. The perſon conceals himſelf in a green arbour, from which he can ſee through a loop-hole a pole, which he has faſtened to a neighbouring tree; the Thruſhes are invited by the call, and expecting to meet with their companions, alight on the pole, and fall by the ſhot of the fowler.

The FIELDFARE*.

La Litorne, Buff.
Turdus Pilaris, Linn. Gmel. Gesner, and Aldrov.
Turdus Pilaris, seu *Turdela*, Briss.

THIS Thrush is the largest after the Missel; and like it can hardly be decoyed by the call, but may be caught by a noose. It differs from the other Thrushes by the yellow colour of its bill, the deeper brown of its legs, and the cinereous sometimes variegated with black, which spreads over its head, behind its neck, and upon its rump.

The male and female have the same cry, which will equally attract the wild Fieldfares in the season of migration †. But the female is distinguished from the male by the colour of her bill, which is much duller. These birds, which breed in Poland and Lower Austria ‡, never nestle in France. They arrive in flocks with the Red-Wing about the beginning of December, and make a loud noise as they fly §.

* In Greek, Τριχας, which is also adopted in Latin: in Italian, *Viscada*, *Viscardo*, (I suspect these names to belong to the Missel Thrush:) in Spanish, *Zorzol*: in German, *Krammet-Vogel*, *Ziemmer*: in Swiss, *Reckolter*, *Wecholter*, *Wachholder-drostel*: in Polish, *Drozd-Srzedni*, *Kwiczot*: in Danish, *Dobbelt Kramsfugl*: in Norwegian, *Graae Trost*, *Field-Trost*, *Norden Vinds Pibe*: in Swedish, *Kramsfogel*, *Snoskata*.

† Frisch. ‡ Klein, and Kramer. § Rzaczynski.

They

They haunt the unploughed fields which are interſperſed with juniper buſhes, and when they appear again in the ſpring *, they prefer the wet meadows. In general they inhabit the woods much leſs than the two preceding ſpecies. Sometimes they make an early but tranſient appearance when the ſervices are ripe, of which they are very fond, though they nevertheleſs return at the uſual time.

It is not an uncommon thing to ſee the Fieldfares aſſemble to the number of two or three thouſand in a ſpot where there are ripe ſervices, which they devour with ſuch voracity, that they throw half of them on the ground. After rains they frequently run along the ditches in ſearch of worms and ſlugs. In the time of hard froſts, they live upon the haws of the white-thorn, the berries of the miſletoe, and thoſe of other plants †.

We may infer then that the Fieldfares are of a much more ſocial diſpoſition than the Throſtles or the Miſſels. They ſometimes go ſingle, but for the moſt part they form, as I have already remarked, very numerous flocks, fly in a body, and ſpread through the meadows in ſearch of food, never loſing ſight of their ſociety. They all collect together upon the ſame tree at certain hours of the day, or when at any time

* They arrive in England about the beginning of October, and depart in the month of May. BRITISH ZOOLOGY.

† Lottinger.

they

they are alarmed at the near approach of a perſon.

Linnæus mentions a Fieldfare, which was bred in the houſe of a wine-merchant, and became ſo familiar that it would run along the table and drink the wine out of the glaſſes; it drank ſo much that it grew bald, but being ſhut up in its cage and denied wine, it recovered its plumage *. This little anecdote preſents two remarkable facts; the effect of wine upon the feathers of a bird, and the inſtance of a tame Fieldfare, which is very uncommon; for the Thruſhes cannot be, as I have before ſaid, eaſily domeſticated.

The Fieldfares are the more numerous in proportion to the ſeverity of the weather; they ſeem to be even a ſign of its continuance, for the fowlers and thoſe who live in the country judge that the winter is not over as long as the Fieldfares are heard. They retire in ſummer into the northern countries, where they breed and find abundance of junipers. Friſch aſcribes to this ſort of food the excellent quality he diſcovered in their fleſh. I own that there is no diſputing about taſtes, but I muſt ſay that in Burgundy this Thruſh is reckoned very indifferent eating, and that in general the flavour communicated by juniper is always ſomewhat bitter. Others aſſert that the fleſh of the Fieldfares is

* *Fauna Suecica*, p. 71.

never

never better or more ſucculent than when it feeds on worms and inſects.

The Fieldfare was known by the ancients under the name of *Turdus Pilaris;* not becauſe it has been always caught with a nooſe, as Salerne ſays, a quality which would not have diſtinguiſhed it from the other Thruſhes, but becauſe the hairs or black briſtles round its bill, which project forwards, are longer in this ſpecies than in the Throſtle or the Miſſel. We may add, that its claws are very ſtrong, as remarked in the Britiſh Zoology. Friſch relates, that if the young of the Miſſel be put in a Fieldfare's neſt, it will feed and educate them as its own; but I would not thence infer, as Friſch has done, that we might expect to obtain an hybridous race; for no perſon ſurely looks for a new breed between the hen and the drake, though the hen often rears whole hatches of ducklings. [A]

[A] Specific character of the Fieldfare, *Turdus Pilaris*:—"Its "tail-quills are black, the outermoſt whitiſh at the tip of their "inner margin, the tail and rump hoary." It is ten inches long, ſeventeen broad, and weighs four ounces. The Fieldfares never breed in Great Britain, but arrive in flocks with the Red-Wings in the end of September, and retire in the beginning of March: but what is ſingular, they appear for a fortnight about Michaelmas, and again for about a week in April. Linnæus and Latham think that the Fieldfares were the Thruſhes which the Romans fattened in their voleries.

VARIETY *of the* FIELDFARE.

The PIED or SPOTTED FIELDFARE. It is variegated with white, black, and many other colours, ſo diſtributed that except the head and the neck, which are white ſpotted with black, and the tail, which is entirely black, the duſky hues, interſperſed with white ſpots, prevail on the upper-part of the body; and, on the contrary, the light colours, eſpecially the white, are ſpread over the lower-part marked with black ſpeckles, moſt of which are ſhaped like ſmall creſcents. This Fieldfare is of the ordinary ſize.

We ought to refer to this the *White-headed Fieldfare* of Briſſon. It has no black ſpeckles, and as its white is what alone diſtinguiſhes it from the common Fieldfare, we may conſider it as intermediate between that and the Spotted Fieldfare. It is even natural to ſuppoſe that the change of plumage would begin at the head, ſince the colour of that part varies in different individuals.

FOREIGN BIRDS,

WHICH ARE RELATED TO THE FIELDFARE.

I.

The CAYENNE FIELDFARE.

Turdus Cayanensis, Gmel.
The Cayenne Thrush, Lath.

I REFER this Thrush to the Fieldfare, because it appears to be more closely related to that species than to any other, by the colour of the upper-part of its body and of its legs. It differs in many respects from the whole genus: its breast and the under-part of its body are not so distinctly dappled; its plumage is more extensively variegated, though in a different manner, almost all the feathers of the upper and underside of the body being edged with a lighter colour, which marks nicely their shape; and lastly, the lower mandible is scalloped near the point;—and these differences are sufficient to constitute it a distinct species, till we are better acquainted with its habits and dispositions *.

* Specific character of the *Turdus Cayanensis*:—"It is cinereous, below partly white, the greater coverts of its wings, and the quills, black; the upper-part of the throat, the bill, and the legs, black." It is of the size of the Throstle, being eight inches long.

II. The

II.

The CANADA FIELDFARE.

Turdus Migratorius, Linn. Gmel. and Klein.
Turdus Canadensis, Briss.
The Fieldfare of Carolina, Catesby.
The Red-breasted Thrush, Penn. and Lath.

Fieldfare is the name which Catesby applies to the Thrush described and figured in his Natural History of Carolina; and I adopt it the more readily, since that species spends at least a part of the year in Sweden, and could thence migrate into the New World, and produce other varieties. In the Canada Fieldfare the orbits are white, there is a spot of the same colour between the eye and the bill, the upper-part of the body is brown, the under orange before, and variegated behind with dirty white and rusty brown, shaded with a greenish tinge; there are also some speckles under its throat, whose ground colour is white. In winter it advances in numerous flocks from the northern parts of America to Virginia and Carolina, and returns in the spring. It resembles our Fieldfare in this circumstance, but it sings better *. Catesby says that it has a

* We must remember that the song of a bird cannot be known unless it be heard in the season of love, and that the Fieldfare never breeds in our climates.

sharp

ſharp note like the Guy Thruſh or Miſſel. He alſo tells us that one of theſe Canada Fieldfares having diſcovered the firſt privet that was planted in Virginia, took ſo great a liking to the fruit, that it remained all the ſummer. Cateſby was informed that theſe birds breed in Maryland, where they remain the whole year. [A]

[A] Specific character of the *Turdus Migratorius* :—" It is " gray, its belly rufous, its eyelids white, the outermoſt tail-quill " white at its inner tip." The Red-breaſted Thruſhes ſeem to traverſe the whole extent of North America. In Hudſon's Bay they appear in pairs about the beginning of May : at Mooſe Fort they neſtle, and hatch in fourteen days; at Severn Settlement, four degrees farther north, they require twenty-ſix. In the State of New-York they arrive in February, lay their eggs in May, and retire ſouthwards in October. They live upon worms, inſects, the ſeeds of the ſaſſafras ſhrub, and various ſorts of berries. Their neſt is compoſed of roots, moſs, &c. The male is aſſiduous in aiding his mate during incubation. She lays four or five eggs, of a fine ſea-green.

M

The

III.

The RED-WING*.

Le Mauvis, Buff.
Turdus Iliacus, Linn. and Gmel.
Turdus Minor, Gesner.
Turdus Illas, seu *Tylas*, Aldrov.
The Red-Wing, *Swinepipe*, or *Wind Thrush*, Will.

This small Thrush is the most useful of them all, since it is the best to eat, especially in Burgundy, where its flesh is delicious †. Besides, it is oftener caught in the noose than any other, and is therefore the most valuable species both for its quantity and its quality ‡. It generally appears the second, that is, after the Throstle and before the Fieldfare; and it arrives in large bodies in November, and departs before Christmas. It breeds in the woods near Dantzic §,

* In Italian, *Malvizzo*, *Tordo-sacello*, *Cion*, *Cipper*: in Spanish, *Malvis*: in German, *Wein-drostel* (Wine-throstle), *Roth-drostel* Red-throstle), *Heide-drostel* (Heath-throstle), *Pfief-drostel* (Pipe-throstle), *Behemle*, *Boemerlin*, *Boemerle*: in Swiss, *Berg-Trostel*, *Wintzel*, *Girerle*, *Gixerle*: in Polish, *Drozd-mnieyssy*: in Swedish, *Klera*, *Kladra*, *Tall-Trast*.

† Linnæus asserts the contrary, *Syst. Nat.* p. 169. This difference between one country and another depends probably on that of the quality of the food, or perhaps on that of tastes.

‡ Frisch and the fowlers assert, that it is not easily taken with nooses, when they are made of white or of black hairs: in Burgundy they are made of these two sorts twisted together.

§ Klein.

 but

but ſeldom or never ſettles in our provinces, or in Lorraine, where it arrives in April, and retires about the end of the ſame month, and appears not again till autumn; though that country affords abundance of proper food in its vaſt foreſts. It halts there a certain time at leaſt, and does not, as Friſch aſſerts, remove merely into ſome parts of Germany. Its common food is berries and ſmall worms, which it finds by ſcraping the ground. It is diſtinguiſhed from the other Thruſhes, by its feathers being more gloſſy and ſhining, its bill and eyes of a deeper black than the Throſtle, whoſe ſize it approaches, and by its having fewer ſpeckles on the breaſt. It is alſo remarkable for the orange colour under its wing, a circumſtance which has occaſioned its being called in ſeveral languages, *Red-winged Thruſh.*

Its ordinary cry is *tan*, *tan*, *kan*, *kan;* and when it perceives a fox, its natural enemy, it leads him off to a great diſtance; as do alſo the Blackbirds, repeating always the ſame notes. Moſt naturaliſts remark that it never ſings; but this aſſertion needs to be qualified, and we can only ſay that it is ſeldom heard to ſing in countries where it does not appear in the ſeaſon of love, as in France, England, &c. An excellent obſerver, M. Hebert, has informed me, that he has witneſſed its chanting in the ſpring in Brie; twelve or fifteen of them ſat on a tree and warbled like linnets. Another obſerver, who lives in in the ſouth of Provence, tells me, that the Red-Wing only whiſtles, which it does inceſſantly; we may infer, therefore, that it does not breed in that country.

Ariſtotle mentions it by the name of *Ilian Thruſh*, as being the ſmalleſt and the leaſt ſpotted of the Thruſhes *. This epithet ſeems to imply that it was brought into Greece from the coaſts of Aſia, where once ſtood *Ilium*, the city of Troy.

I have traced an analogy between this ſpecies and the Fieldfare. They are both foreign, and only viſit our climate twice a year †; they aſſemble in numerous flocks at certain hours to chirp together; they are ſimilarly marked with ſpeckles on the breaſt. But the Red-Wing is alſo related to the Throſtle; its fleſh is not inferior in quality, the under-ſurface of its wing is yellow, but more lively indeed, and of an orange tinge; it often occurs ſingle in the woods, and viſits the vineyards, like the Throſtle, with which Lottinger has obſerved it often to fly in company, eſpecially in the ſpring. From the whole it appears that this ſpecies is furniſhed with the means of ſubſiſting of the other two,

* Ariſtotle, *Hiſt. Anim.* lib. ix. 20.

† In natural Hiſtory, as in many other ſubjects, general obſervations always admit of exceptions. Though for the moſt part the Red Wing does not ſpend the winter in our climates, I am aſſured by M. Hebert, that he killed one year in a hard froſt ſeveral dozens on a white-thorn, which was ſtill loaded with its berries.

and that in many reſpects it may be regarded as forming the ſhade between the Throſtle and the Fieldfare. [A]

[A] Specific character of the *Turdus Iliacus*:—" Its wings are " ferruginous, its eye-lids whitiſh." It is ſmaller than the Fieldfare, and nearly nine inches long. It breeds in the north of Europe, in hedges and buſhes, and lays ſix eggs of a bluiſh green, ſpotted with black.

FOREIGN BIRDS,

WHICH ARE RELATED TO THE THRUSHES AND BLACKBIRDS.

I.

The BARBARY THRUSH*.

La Grive Baſſette de Barbarie, Buff.
Turdus Barbaricus, Gmel.
The Greek Thruſh, Shaw.

IT reſembles the Thruſhes in its general ſhape, in its bill, and the ſtreaks on its breaſt diſperſed regularly upon a white ground; in ſhort, by all the exterior characters, except its legs and its wings. Its legs are not only ſhorter, but ſtronger; in which it is oppoſite to the *Hoamy*, and ſeems to reſemble ſomewhat our Miſſel, which has its legs ſhorter in proportion than the other three ſpecies. With regard to the plumage, it is extremely beautiful: the prevailing colour on the upper-part of the body, including the head and the tail, is a light brilliant green, and the rump is tinged with a fine yellow, as alſo the extremity of the coverts of the tail and

* Termed *Baſſette,* on account of its ſhort legs.

of the wings, of which the quills are of a leſs vivid colour. But this enumeration of the colours, were it even more complete, will by no means give a juſt idea of the effect which they produce in the bird itſelf; a pencil, and not words, can exhibit its beauty. Dr. Shaw, who ſaw this Thruſh in its native country, compares its plumage to that of the richeſt birds of America; he adds, that it is not very common, and appears only in the ſeaſon when the figs are ripe. This would ſhew that theſe fruits direct its migration, and in this ſingle fact I perceive two analogies betweeen this bird and the Thruſhes; both birds of paſſage, and both exceſſively fond of figs *.

* Specific character of the *Turdus Barbaricus*:—"It is green, "its breaſt ſpotted with white; its rump, and the tip of its tail, "yellow."

II.

The RED-LEGGED THRUSH.

Le Tilly, ou *La Grive Cendree d'Amerique*, Buff.
Turdus Plumbeus, Linn. and Gmel.
Merula Americana Cinerea, Briſſ.
Turdus Thilius, Molin.
Merula Tilli, Feuillée.

All the upper-part of the body of this bird, its head and neck, are of a deep aſh-colour; which

which extends over the ſmall coverts of the wings, and, paſſing under the body, riſes on the one hand as far as the throat, without ſuffering any change; and, on the other, deſcends to the lower belly, ſhading however gradually into white, which is alſo the colour of the coverts under the tail. The throat too is white, but dappled with black; the quills and the great coverts of the wings are blackiſh, and edged exteriorly with cinereous. The twelve quills of the tail are tapered and blackiſh, like thoſe of the wing, but the three outer ones on each ſide are terminated by a white ſpot, which is the larger, the nearer it is to the margin. The iris, the orbits, the bill, and the legs, are red; the ſpace between the eye and the bill black, and the palate tinged with a vivid orange.

The total length is about ten inches; its alar extent near fourteen, its tail four, its leg eighteen lines, its bill twelve, its weight two ounces and a half; laſtly, its wings when cloſed do not reach the middle of the tail.

This bird is ſubject to variety; for in the one obſerved by Cateſby, the bill and throat were black. May we not aſcribe this difference of colours to ſex? Cateſby only ſays that the male is a third ſmaller than the female; he adds, that theſe birds feed on the berries of the tree which produces gum elemi.

It is found in Carolina, and, according to Briſſon, it is very common in the iſlands of Andros and Ilathera.

III.

The SMALL THRUSH of the Philippines.

Turdus Philippenſis, Gmel.
The Philippine Thruſh, Lath.

We owe this to Sonnerat. The fore-part of its neck and breaſt are dappled with white upon a rufous ground; the reſt of the lower-part of the body is dirty white, bordering on yellow, and the upper-part of the body is of a deep brown, with an olive tinge.

The ſize of this bird is inferior to that of the Red-Wing; we cannot aſcertain its alar extent, ſince the wing-quills in the ſpecimen which we have obſerved are incomplete.

IV.

The HOAMY of China.

Turdus Sinenſis, Linn. and Gmel.
The Chineſe Thruſh, Lath.

Briſſon is the firſt who has deſcribed this bird, or rather the female of it. This female is ſome-

what

what ſmaller than a Red-Wing, which it reſembles as well as the Throſtle, and ſtill more the Canada Thruſh, its legs being proportionably longer than in the other Thruſhes; they are yellowiſh, as is the bill; the upper-part of the body is of a brown, bordering upon rufous, the under of a light and uniform rufous; the head and neck are ſtriped longitudinally with brown; the tail is alſo of the ſame colour, only barred tranſverſely.

Such is nearly the deſcription of the external appearance of this bird; but we are not informed with regard to its inſtincts and habits. If it be really a Thruſh, as it is ſaid, its breaſt is like that of the Red Thruſh, not dappled.

V.

The LITTLE THRUSH of St. Domingo.

This Thruſh is, in point of ſmallneſs, like the American Thruſh; its head is ornamented with a ſort of crown or cap of bright orange, verging upon red.

The ſpecimen figured by Edwards, Pl. 252, differs from ours in not being dappled under the belly. It was caught in November 1751, at ſea, eight or ten leagues off the iſland of St. Domingo; which led Edwards to ſuppoſe that it was one of thoſe birds of paſſage which every year

year leave the continent of North America on the approach of winter, and depart from the Cape of Florida in queſt of milder ſeaſons. This conjecture was verified. Bartram informed Edwards, that theſe birds arrived in Pennſylvania in the month of April, and remained there during the whole ſummer. He added, that the female built its neſt on the ground, or rather in heaps of dry leaves, where it formed a ſort of excavation; that it lined it with graſs, and always choſe the ſlope of a hill facing the ſouth, and that it layed four or five eggs ſpotted with brown. Such differences in the colour of the eggs, in that of the plumage, and in the mode of neſtling, ſeem to point at a nature diſtinct from that of our European Thruſhes.

VI.

The LITTLE CRESTED OUZEL of China.

I place this bird between the Thruſhes and the Ouzels, or Blackbirds, becauſe it has the port and the ground colours of the Thruſhes, but without the dapples, which we may conſider in general as the diſtinctive character of that genus. The feathers on the crown of the head are longer than the reſt, and the bird can erect them into a creſt. It has a roſe-colour mark

mark behind the eye; another more confiderable of the fame colour, but not fo bright, under the tail; and its legs are of a reddifh brown. Its fize is nearly that of the Lark, and its wings, which when difplayed extend ten inches, fcarcely reach, when clofed, to the middle of the tail. The tail confifts of twelve tapered quills. Brown more or lefs deep is the prevailing colour of the upper-part of the body, including the wings, the creft, and the head, but the four lateral quills on either fide of the tail are tipt with white. The under-part of the body is of this laft colour, with fome tints of brown over the breaft. I muft not forget two blackifh ftreaks, which, rifing from the corners of the bill, and extending over a white ground, make a kind of muftachoe, which has a remarkable effect.

The MOCKING BIRDS.

Les Moqueurs, Buff.

EVERY remarkable bird has always many names, but if it be at the ſame time a native of a foreign climate, this embarraſſing multitude, diſgraceful to Natural Hiſtory, is increaſed by the confuſion of ſpecies. Such is the caſe with the American Mocking Birds. It is eaſy to perceive that Briſſon's Mocking Bird, and the Cinereous Blackbird of St. Domingo, *Pl. Enl.* No. 558, both belong to the ſame ſpecies, the only difference being that the former has ſomewhat leſs of the gray colour in the under-ſide of the body than the latter. It will alſo appear, from compariſon, that Briſſon's Blackbird of St. Domingo is likewiſe the ſame, diſtinguiſhed only by ſome lighter or deeper tints on its plumage and its tail-quills, which are hardly at all tapered. In like manner we ſhall find that the *Tzonpan* of Fernandez is either the female of the *Cencontlatolli*, that is, of the Mocking Bird, as Fernandez himſelf ſuſpects, or at leaſt a permanent variety of that ſpecies *.

* *Hiſtoria Avium Novæ Hiſpaniæ*, cap. xx.—Nieremberg calls it *Tzanpan*, Hiſt. Nat. lib. x. cap. 77.; and Edwards, *Tzaupan*.

It

It is true that its plumage is leſs uniform, being mixed above with white, black, and brown, and below with white, black, and cinereous; but the fundamental colour is the ſame, as alſo its ſize, its general ſhape, its ſong, and climate. We may ſay the ſame of the *Tetzonpan* and *Centzonpantli* of Fernandez *; for in the ſhort mention which that author has made of it, features of analogy are to be met with in ſize, in colour, and in ſong, and no inſtance of diſparity occurs. Beſides, the reſemblance between the names *Tzonpan*, *Tetzonpan*, *Centzonpantli*, ſeems to ſhew that they mark a ſingle ſpecies, and that the diverſity has ariſen from the miſtake of the tranſcriber, or the difference of the Mexican dialects.—Laſtly, we can ſcarcely heſitate to admit among the ſpecies the bird, called by Briſſon the *Great Mocking Bird*, and which he ſays is the ſame with Sloane's Mocking Bird, though, according to the dimenſions given by Sloane, this is the ſmalleſt of the kind; but Sloane regards it as the *Cencontlatolli* of Fernandez, which Briſſon makes his ordinary Mocking Bird. But Briſſon has himſelf, without perceiving it, admitted the poſition which I hold; for he quotes two paſſages from Ray, which applied to the ſame bird, and refers one to his great, and the other to his ſmall ſpecies. The only difference between the two is, that the great Mocking Bird has a

* Cap. cxv.

ſome-

ſomewhat browner plumage, and longer legs *; and its deſcribers have taken no notice of its tapered tail.

After this reduction, there remains only two ſpecies of Mocking Birds, *viz.* The French Mocking Bird and the Ordinary Mocking Bird. I ſhall treat of them in the order I have named them, as it is nearly that of their relation to the Thruſhes.

The FRENCH MOCKING BIRD

Turdus Rufus, Linn. Gmel. and Klein.
Turdus Carolinenſis, Briſſ.
Fox-coloured Thruſh, Cateſby, and Penn.
The Ground Mocking Bird, Lawſon.
The Ferruginous Thruſh, Lath.

None of the American Mocking Birds reſembles ſo much our Thruſhes in the ſpeckles on the breaſt, as this; but it differs widely from them in the proportions between the tail and wings, theſe ending, when cloſed, almoſt where the tail begins. The tail is more than four inches long, which exceeds the third of the whole length of the bird, that being only eleven inches.

* The expreſſion of Sloane is ſomewhat ambiguous; he ſays that its legs and feet are an inch and three quarters long. Does he mean the leg with the tarſus, or the tarſus with the toes? Briſſon underſtood it to ſignify the tarſus alone. (This laſt is certainly the uſual meaning of the *leg* of a bird in Engliſh, and accordingly we have tranſlated the French *pied* by *leg*. *Tr.*)

inches. Its ſize is intermediate between that of the Miſſel and the Fieldfare. Its eyes are yellow, its bill blackiſh, its legs brown, and all the upper-part of the body of a fox colour, but with a mixture of brown. Theſe two colours alſo predominate on the wing-quills, though ſeparately; the rufous on the outer webs, the brown on the inner. The great and middle coverts of the wings are tipt with white, which forms two ſtreaks that croſs the wings obliquely.

The under-ſide of the body is dirty white, ſpotted with a duſky brown, but theſe ſpots are more ſtraggling than in our Thruſhes: the tail is tapered ſomewhat drooping, and entirely rufous. The ſong of the French Mocking Bird has ſome variety, but not comparable to that of the proper Mocking Bird.

It feeds commonly on a kind of black cherries, which are very different from thoſe of Europe, ſince they hang in cluſters. It remains all the year in Carolina and Virginia, and conſequently is not, at leaſt in thoſe provinces, a bird of paſſage:—another analogous circumſtance to our Thruſhes *.

* Specific character of the *Turdus Rufus*:—" It is rufous, be-" low partly ſpotted with white, its wing-quills of an uniform co-" lour, its tail rounded and rufous." It inhabits North America; appears in New-York in May, and retires ſouthwards in Auguſt. It builds in low buſhes, and lays five eggs, thick ſpotted with ferruginous. It is much inferior in the variety of its notes to the true Mocking Bird.

The MOCKING BIRD.

Turdus Polyglottus, Linn. and Gmel.
Mimus Major, Briſſ.
The American Nightingale, Song Thruſh, or *Gray Mocking Bird*, Sloane.
The Mocking Bird, Cateſby.
The Mimic Thruſh, Penn. and Lath.

We have here a ſtriking exception to the general remark made by travellers, that in proportion as the plumage of the birds in the New World are rich, elegant, and ſplendid, ſo their notes are harſh, raucous, and monotonous. The Mocking Bird is, on the contrary, if we believe Fernandez, Nieremberg, and the native Americans, the ſweeteſt choriſter of the feathered race, not excepting the Nightingale. It equals that charming bird in the melody of its ſong; but it poſſeſſes beſides the power of imitating the cries of other animals: hence is derived its name. Nor is it ſatisfied with barely re-echoing the ſounds. It gives them ſoftneſs and grace. Accordingly the ſavages have beſtowed upon it the appellation of *cencontlatolli*; that is, four hundred languages; and the learned have employed the epithet *polyglot*. But the Mocking Bird mingles action with its ſong, and its meaſured movements accompany and expreſs the ſucceſſion of its emotions. Its prelude is to riſe ſlowly with expanded wings, and ſoon ſink

ſink back to the ſame ſpot, its head hanging downwards. Its action now correſponds with the varied nature of its muſic. If the notes are briſk and lively, it deſcribes in the air a number of circles croſſing each other; or it aſcends and deſcends continually in a ſpiral line. If they are loud and rapid, it with equal briſkneſs flaps its wings. Is its ſong unequal? it flutters, it bounds. Do its tones ſoften by degrees, melt into tender ſtrains, and die away in a pauſe more charming than the ſweeteſt melody? it gently diminiſhes its action, glides ſmoothly above its tree, till the wavings of its wings begin to be imperceptible, at leaſt ceaſe, and the bird remains ſuſpended and motionleſs in the air.

The plumage of this American Nightingale by no means correſponds with the beauty of its ſong; the colours are very ordinary, and have neither brilliancy nor variety. The upper-ſide of the body is a grayiſh brown; the upper-ſide of the wings and of the tail are ſtill of a deeper brown; only it is interrupted, 1. on the wings by a white mark which croſſes it obliquely near the middle of its length, and ſometimes a few ſmall white ſpeckles are ſcattered on the fore-part. 2. On the tail by an edging of the ſame white colour; and laſtly, on the head with a circle of the ſame, which forms a ſort of crown*, and extending over the eyes appears like two diſtinct eye-lids. The under-ſide of the body

* Fernandez.

is white from the throat as far as the end of the tail. We perceive in the figure given by Edwards a few dapples, ſome on the ſides of the neck, and others on the white of the great coverts under the wings.

The Mocking Bird approaches the Red-Wing in ſize; its tail is ſomewhat tapered *, and its feet are blackiſh; its bill is of the ſame colour, and with long briſtles that grow above the angles of its opening; laſtly, its wings are ſhorter than thoſe of our Thruſhes, but longer than thoſe of the French Mocking Bird.

It is found in Carolina, Jamaica, New Spain, &c. It in general loves the hot climates, but can ſubſiſt in the temperate. In Jamaica it is very common in the ſavannas of the woody parts of the iſland; it perches on the higheſt branches, and chants its ſong. It often builds its neſt on the ebony trees. Its eggs are ſpotted with brown. It feeds on cherries and the berries of the whitethorn and cornel tree †, and even on inſects. Its fleſh is eſteemed excellent. It is not eaſily raiſed in a cage; but this may be accompliſhed by care and kind treatment. It is beſides a familiar bird, which ſeems to be fond of man, approaches his dwellings, and even perches on the chimney tops.

In the ſubject which Sloane diſſected, the ſtomach was a little muſcular, the liver whitiſh, and

* This appears not in Sloane's figure.

† The Dogwood, *Cornus Florida*, Linn.

the inteftines were folded in a vaft number of circumvolutions. [A]

[A] Specific character of the American Mocking Bird, *Turdus Polyglottus*, LINN. "It is of a dull afh-colour, below pale-afh, the "primary wing-quills white in their outer half." In the northern provinces of America, as New-York, they appear only during the fummer. In the warmer parts of the continent they fing inceffantly, by night and day, from March to Auguft. They often breed in fruit-trees, but are fhy during the time of incubation. They lay five or fix eggs, which are blue, thickly fpotted with rufous.

There is a fmaller kind of Mocking Bird, which has obtained the following appellations from naturalifts:

Turdus Orpheus, Linn. and Gmel.
Mimus, Briff.
Avis Polyglotta, Will.
The Leffer Mocking Bird, Edw.
The Mocking Thrufh, Lath.

Its fpecific character:—"Its back is dufky, its breaft and its "lateral tail-quills are whitifh, its eye-brows white."

M

The BLACKBIRD*.

Le Merle, Buff.
Turdus Merula, Linn. and Gmel.
Merula, Belon, Gesner, Briss. &c.

THE adult male of this species is of a deeper and purer black than the Raven, and hence its English name. Indeed, except the orbits, the talons, and the sole of the foot, which have always a yellow cast, it is entirely black in every position. In the female, on the contrary, the same decided black is not spread through the whole of the plumage; it is mixed with different shades of brown, ferruginous, and gray, the bill is but seldom yellow, and the song is different from that of the male;—all these circumstances combined have made it be mistaken for a bird of another species.

The Blackbirds are distinguished from the Thrushes not only by the colour of their plumage, and the different livery of the male and of the female, but by their song and their natural habits. They never fly in flocks like the

* In Greek, Κοσσυφος, or Κοτlυφος, also Κοψυχος: the Latin name *Merula* is derived by Varro from *Mera*, (*mere, alone, solitary,*) which denotes the retired disposition of the bird: in Italian, *Merlo*: in Spanish, *Mierla*: in Portuguese, *Melroa*: in German, *Schwartze Amsel* (Black Ouzel): in Flemish, *Merlaer*, *Meerel*: in Swedish, *Kohl-trost*: in Danish and Norwegian, *Solsort*.

 Thrushes,

No. 76

THE BLACKBIRD

Thruſhes, and though they are more ſavage with regard to each other, they are leſs ſo with regard to man; for they are more eaſily tamed, and live nearer the hamlets. They are alſo reckoned very crafty, becauſe they have a quick ſight to deſcry the fowler at a diſtance, and ſhun his approach. But if we ſtudy their nature more cloſely, we ſhall find that they are more reſtleſs than cunning, more timorous than ſuſpicious, ſince they allow themſelves to be caught with bird-lime, with nooſes, and with all ſorts of ſnares, provided the hand which ſets theſe be concealed.

When they are ſhut up with other birds, their natural inquietude degenerates into petulance; they purſue and continually haraſs their companions in ſlavery; and for that reaſon they cannot be admitted into voleries, where ſeveral kinds of ſmall birds are kept.

They may be raiſed apart for the ſake of their ſong; not indeed on account of their natural muſic, which is hardly tolerable except in the fields, but becauſe they have a facility of improving their notes and of learning others, of imitating tones of different inſtruments *, and even the human voice †.

As the Blackbirds, like the Thruſhes, early engage in love, they ſoon begin to warble; and

* Olina, *Uccelliera*, p. 29.

† Olina. *Ibid.*—Philoſtrat. *Vita Apollonii*, lib. vii.—Geſner, *De Avibus*, p. 606.

as they have more than one hatch, they chant before the vernal warmth, and continue their ſtrain when the other ſongſters of the grove droop in ſilence. This circumſtance has led ſome to fancy that they never ſuffer any change of plumage; but ſuch a ſuppoſition is neither true nor probable *. They are found in the woods, towards the end of the ſummer, in moult, ſome having their head entirely bald: Olina and the author of the Britiſh Zoology ſay, that, like the other birds, it is ſilent during that time; the latter adds, that ſometimes it renews its ſong in the beginning of winter, but moſt commonly it has in that ſeaſon only a harſh diſcordant cry.

The ancients pretended that in winter its plumage changed into rufous †; and Olina, one of the moderns the beſt acquainted with the birds which he deſcribes, ſays, that this happens in autumn: whether it be becauſe this alteration of colour is the effect of moulting, or that the females and the young Blackbirds, which really incline to rufous rather than to black, are then more numerous and oftener ſeen than the adult males.

Theſe birds make their firſt hatch in the end of winter; it conſiſts of five or ſix eggs of a bluiſh green, with frequent and indiſtinct ſpots

* "The blackbirds, thruſhes, and ſtares, loſe not their feathers." PLINY, lib. x. 24.

† Ibid, lib. x. 29.

of

of a ruſt colour. This firſt hatch ſeldom proſpers, becauſe of the ſeverity of the weather; the ſecond ſucceeds better, though it is only of four or five eggs. The Blackbirds neſt is conſtructed nearly like that of the Thruſhes, except that it is lined with a matting. It is commonly placed in buſhes or trees of a moderate height. They would ſeem naturally diſpoſed to place it near the ground; and experience alone of the danger of ſuch a ſituation inſtructs them to give it ſome elevation *. A neſt was brought to me only once, which had been found in the trunk of a hollow pear-tree.

Moſs, which always occurs on the trunk, and mud, which is to be found at the foot of the tree, or in its neighbourhood, are the materials that form the body of the neſt. Stalks of graſs and ſmall roots are the ſofter ſubſtances with which they line it; and they labour with ſuch aſſiduity that in eight days they complete the work. The female hatches alone, and the male is no farther concerned than in providing her ſubſiſtence. The Author of the *Treatiſe on the Nightingale* affirms, that he has ſeen a young Blackbird of the ſame year, but already ſtrong, cheerfully engage in rearing the infant

* " I have diligently examined a neſt found near the ground " in a place overgrown with thorns." GESNER.—A Blackbird obſerving that a cat ate its two firſt hatches in its neſt at the bottom of a hedge, made a third on an apple-tree, eight feet high. SALERNE.

 brood

brood of its own ſpecies; but he does not inform us of its ſex.

I have remarked that the young drop their feathers more than once the firſt year; and that, at each time, the plumage of the male becomes blacker, and the bill yellower, beginning at its baſe. With regard to the females, they retain, as I have ſaid, the colours of infancy, as they alſo retain moſt of its qualities. However, the inſide of their mouth and throat is yellow, like the males; and in both may be perceived a frequent motion of the tail upwards and downwards, with a ſlight ſhudder of the wings, accompanied by a feeble broken cry.

Theſe birds do not leave the country in winter *, but chooſe ſituations the beſt ſheltered, ſettling commonly in the thickeſt woods, eſpecially when theſe are ſupplied with perennial ſprings, and conſiſt of evergreens, ſuch as pines, firs, laurels, cypreſſes, myrtles, junipers, which both afford them ſubſiſtence, and protect them from the rigour of the ſeaſon. They ſometimes ſeek for cover and food in our gardens.

* Many people aſſert that they leave Corſica about the 15th of February, and return not till October; but M. Artier, Royal Profeſſor of Philoſophy at Baſtia, doubts the fact, becauſe in that iſland they can always enjoy the proper temperature; in cold weather in the plains, and in the ſultry ſeaſons in the mountains. He adds, that there is always abundance of food, berries of all kinds, grapes, and particularly olives, which in Corſica are not entirely gathered till the end of April. Lottinger believes that the males paſs the winter in Lorraine, but that the females remove to a little diſtance in ſevere weather.

The

The wild Blackbirds feed on all ſorts of berries, fruits, and inſects; and as no country is ſo ſterile as not to afford ſome of theſe, and as the Blackbird is reconciled to all climates, it is found in almoſt every part of the world, but varying according to the impreſſions which it receives.

Thoſe which are kept in the cage, eat fleſh alſo, either dreſſed or minced, bread, &c. but it is ſaid that the kernels of pomegranates prove poiſonous to them as to the Thruſhes. They are very fond of bathing, and they muſt have plenty of water in the voleries. Their fleſh is good, and not inferior to that of the Miſſel or the Fieldfare, and ſeems even to be preferred to that of the Throſtle and of the Red-Wing, in countries where it can require a ſucculence from the olives, and a perfume from the myrtle-berries. The birds of prey are as fond of feaſting on them as man, and commit an equal havoc: without that their multiplication would be exceſſive. Olina fixes their period of life at ſeven or eight years.

I diſſected a female, which was taken on its eggs about the 15th of May, and which weighed two ounces and two gros. In the *ovarium* was a cluſter conſiſting of a great number of unequal ſized eggs; the largeſt two lines in diameter, and of an orange colour; the ſmalleſt were of a lighter colour, and of a ſubſtance leſs opaque, and about one-third of a line in diameter. Its

bill

bill was quite yellow, alſo the tongue and the whole inſide of the mouth, the inteſtinal tube ſeventeen or eighteen inches long, the gizzard very muſcular, and preceded by a bag formed by the dilatation of the *œſophagus;* the gall bladder oblong, and the *cæcum* wanting. [A]

[A] Specific character of the Blackbird, *Turdus Merula*:—"It "is black, its bill and eyelids are yellow." It builds earlier than any other bird; its neſt is formed with moſs, withered graſs, leaves, &c. lined with clay, over which is ſpread ſome hay.

M

VARIETIES *of the* BLACKBIRD.

THOSE THAT ARE WHITE, OR SPOTTED WITH WHITE.

The plumage of the Blackbird is ſubject, like that of the Raven, the Crow, the Jackdaw, and other birds, to great changes, from the influence of the climate, or from the action of leſs obvious cauſes. In fact, white ſeems to be in moſt animals, what it is in many plants, the colour into which all the others, and even the black, degenerate by a quick tranſition, and without paſſing through the intermediate ſhades.

The only varieties of this ſort which appear to belong to the common Blackbird, are, 1. the White one, which was ſent to Aldrovandus at Rome; and, 2. the White-headed one of the ſame author. Both theſe have the yellow bill and feet of the ordinary ſpecies.

N.º 77

THE RING-OUZEL.

The RING OUZEL*.

Le Merle a Plastron Blanc, Buff.
Turdus Torquatus, Linn. and Gmel.
Merula Torquata, Briss. Ray, and Will.

THIS species is marked above the breast with a horse-shoe, which, in the male, is of a very bright white, but in the female is of a dirty tawny colour; and as the rest of the female's plumage is a rufous brown, the horse-shoe appears much less distinct, and is sometimes entirely obscured †. Hence some nomenclators have imagined that the female belonged to a particular species, which they termed *The Mountain Blackbird*.

The Ring Ouzel much resembles the common Blackbird; the ground colour of their plumage is black, the corners and the inside of their bill yellow; they are nearly of the same size and the same port: but the former distinguished by the horse-shoe, by the white enamel of its plumage, chiefly on the breast, belly, and wings ‡; by its bill, which is shorter and not

* In Italian, *Merula Alpestro* (or Crag Blackbird): in German, *Ring-Amsel, Rotz-Amsel* (snotty, or filthy Ouzel, because it feeds sometimes on the maggots found in horse-dung): *Wald Amsel* (Wood Ouzel): *Stein-Amsel, Berg-Amsel* (Mountain Ouzel): *Schnee-Amsel* (Snow Ouzel).

† WILLUGHBY.

‡ Willughby saw at Rome one of these birds, which had its horse-shoe gray, and all its feathers edged with the same colour. He judged it was a young bird, or a female.

so

ſo yellow; by the ſhape of the middle-quills of the wings, which are ſquare at the end, with a ſmall-projecting point in the centre, formed by the extremity of the ſhaft; laſtly, by its cry*, which is different, as alſo its habits and diſpoſitions. It is a real bird of paſſage, though its route cannot be preciſely traced. It follows the chain of the mountains, but does not keep in any certain track †. It ſeldom appears in the neighbourhood of Montbard, except in the beginning of October, when it arrives in ſmall bodies of twelve or fifteen, and never in larger numbers. Theſe ſeem to be a few families that have ſtraggled from the great body; they ſeldom ſtay more than two or three weeks, and on the ſlighteſt froſt entirely diſappear. But I muſt own that Klein informs us that theſe birds were brought to him alive in winter. They repaſs about April or May, at leaſt in Burgundy, Brie ‡, and even in Sileſia and in Friſia, according to Geſner.

It is uncommon for the Ring Ouzels to inhabit the plains in the temperate part of Europe;

* This cry in autumn is *crr, crr, crr:* but a perſon of veracity aſſured Geſner, that he heard this Ouzel ſing in the ſpring, and very agreeably.

† It does not appear every year in Sileſia, according to Schwenckfeld: this is alſo the caſe in certain cantons of Burgundy.

‡ M. Hebert aſſures me that in Brie, where he has fowled much at all ſeaſons, he killed a great number of theſe Ouzels in the months of April and May, and that he never chanced to meet with any in the month of October. In Burgundy, on the contrary, they ſeem leſs rare in autumn than in ſpring.

yet

yet Salerne affirms that their nests have been found in Sologne and in the forest of Orleans; that these nests were not constructed like those of the ordinary Blackbird; that they contained five eggs of the same size and colour (a circumstance different from what happens in the Blackbirds); that these birds breed in the ground at the foot of bushes, and hence probably they are called *Bush-Birds* or *Terrier Blackbirds* *. Certain it is that in some seasons of the year they are very frequent on the lofty mountains of Sweden, of Scotland, of Auvergne, of Savoy, of Switzerland, of Greece, &c. It is even probable that they are spread in Asia, and in Africa as far as the Azores; for this species, so social, so fond of dwelling in mountains, and having its plumage marked with white, corresponds well to what Tavernier says of the flocks of Blackbirds which pass from time to time on the frontiers of Media and Armenia, and rid the country of grasshoppers. It also agrees with the account which Adanson gives of those Blackbirds spotted with white, which he saw on the summits of the mountains in the island of Fayal, keeping in flocks among the arbutes, on the fruit of which they fed, chattering continually †.

Those which ramble in Europe subsist likewise on berries. Willughby found in their sto-

* *Merles Terriers*, ou *Bouissenniers*.

† Voyage au Senegal.

mach

mach veſtiges of inſects, and berries reſembling gooſeberries; but they prefer thoſe of ivy and grapes. It is in the ſeaſon of vintage that they are generally ſo fat, and their fleſh ſo ſavoury and ſucculent.

Some fowlers ſay that the Ring Ouzels attract the Thruſhes; they remark too that they allow themſelves to be more eaſily approached than the common Blackbirds, though they are more difficult to decoy into ſnares.

I found, on diſſection, that their gall-bladder is oblong, very ſmall, and conſequently quite different from what Willughby deſcribes it to be; but the ſituation and form of the ſoft parts, it is well known, are very ſubject to vary in animals. The ventricle was muſcular, its inner coat wrinkled as uſual, and inadheſive. In this membrane I ſaw fragments of juniper berries and nothing elſe. The inteſtinal canal, meaſured between its two extreme orifices, was about twenty inches; the ventricle or gizzard was placed between the fourth and fifth of its length. Laſtly, I perceived ſome traces of *cæcum*, of which one appeared to be double. [A]

[A] Specific character of the Ring-Ouzel, *Turdus Torquatus*, LINN.—"It is blackiſh, with a white collar, its bill yellowiſh." It is larger than the Blackbird, being eleven inches long and ſeventeen broad. It inhabits the mountainous parts of this iſland in ſmall bodies of five or ſix.

The *Merula Saxatilis*, or the *Rock Ouzel*, is reckoned by Latham the young of the Ring-Ouzel, from which it differs, chiefly by the dulneſs of its colours.

VARIETIES *of the* RING-OUZEL.

I.

THOSE WHICH ARE WHITE, OR SPOTTED WITH WHITE.

Ariſtotle was acquainted with White Ouzels, and made them a diſtinct ſpecies, though they have the ſame ſong and the ſame bulk with the common Ouzel or Blackbird; but he knew that their inſtincts were different, ſince they preferred the mountains *: and theſe are the only diſtinctive characters which Belon admits †. They are found not only in the mountains of Arcadia, of Savoy, and of Auvergne, but alſo in thoſe of Sileſia, and among the Alps and Appennines, &c. ‡. They are alſo birds of paſſage, and migrate with the Ring-Ouzel at the ſame ſeaſon. The white colour of the horſeſhoe in the Ring-Ouzel may extend over the reſt of the plumage. I ſhould therefore conceive that theſe, though uſually referred to the Blackbirds, belong really to the Ring-Ouzels. In the white one which I obſerved, the quills of the wings and tail were whiter than any of the reſt, and the upper-part of the body, except

* "They are frequent about Cyllene in Arcadia, and breed "no where elſe." *Hiſt. Anim.* lib. ix. 19.

† He ſays expreſsly that the White Ouzel never deſcends into the plains.

‡ WILLUGHBY.

the

the top of the head, was of a lighter gray than the under. The bill was brown, with a little yellow on the edges; there was alſo yellow under the throat and on the breaſt, and the legs were of a deep gray brown. It was caught in the vicinity of Montbard in the beginning of November before the froſt; that is, at the exact time of the paſſage of the Ring-Ouzel; for a few days before, two of that ſpecies were brought to me.

In thoſe which are ſpotted, the white is combined variouſly with the black; ſometimes it is confined to the quills of the wings and tail, which are commonly ſuppoſed to be leaſt ſubject to change of colour *; ſometimes it forms a collar that encircles the neck, but is not ſo broad as the white horſe-ſhoe of the Ring-Ouzel. This variety did not eſcape Belon, who ſays that he ſaw in Greece, in Savoy, and in the valley of Maurienne, a great number of collared Blackbirds, ſo called on account of a white line which bent quite round the neck. Lottinger, who had an opportunity of obſerving theſe birds in the mountains of Lorraine, where they ſometimes breed, informs me, that they commence breeding very early; that they conſtruct and place their neſt nearly like the Thruſh; that the education of their young is completed before the end of June; that they retire every year, but that the time of their departure is not fixed; that this uſually be-

* Aldrovandus.

gins

gins about the end of July, and lafts the whole of Auguft, during which time not one is feen in the plain, a proof that they follow the chain of the mountains, but their retreat is uncertain. Lottinger adds, that this bird, which formerly was very common in the Vofges, is now feldom found there.

II. The GREAT MOUNTAIN OUZEL.

It is fpotted with white, has no horfe-fhoe, and is larger than the Miffel. It arrives in Lorraine about the end of autumn, and is then exceffively fat. The bird-catchers feldom fucceed with it; it feeds upon fnails, and is dexterous in breaking the fhells. When thefe fail, it fubfifts on ivy-berries. It is excellent eating; its ftrains, far inferior to thofe of the Blackbird, are harfh and difmal *.

* I am indebted for thefe facts to Dr. Lottinger.

The ROSE-COLOURED OUZEL *.

Le Merle Couleur de Rose, Buff.
Turdus Roseus, Linn. and Gmel.
Merula Rosea, Briss. Ray, and Will.
Sturnus Roseus, Scopoli.
The Rose, or *Carnation-coloured Ouzel*, Penn.

ALL the ornithologists, who have taken notice of this bird, mention it as very rare, as foreign, and little known; that it is seen only in its passage, and the country to which it belongs is uncertain. Linnæus tells us indeed, that it inhabits Lapland and Switzerland; but he says nothing with regard to its instincts and mode of life. Aldrovandus, who first described this bird, only remarks that it appears sometimes in the plains near Bologna, where it is known by the birdcatchers under the name of *Sea-Stare*, (*Storno Marino*); that it sits on the dung-hills, grows very fat, and is excellent eating. Two birds of this kind were found in England, and Edwards supposes that they were driven thither by the violence of the wind. We have observed several in Burgundy, which had been caught in their passage, and it is probable that they pursue their excursions as far as Spain, if what Klein says be

* In Spanish, *Tordos*: in German, *Haarkopfige-Drossel* (Hair-headed Thrush).

true,

N.º 78

THE ROSE COLOURED THRUSH.

true, that they have a name in the Spaniſh language.

The plumage of the male is remarkable; its head and neck, and the quills of its wings and of its tail, are black, with brilliant reflexions which play between green and purple. The belly, the back, the rump, and the ſmall coverts of the wings are of a roſe colour, which has two tints, the one light, the other deep, with a few black ſpots ſcattered here and there on a kind of ſcapulary, which deſcends above as far as the tail, and below to the abdomen. Beſides, its head is ornamented with a ſort of creſt which reclines like that of the Chatterer, and which muſt have a fine effect when the bird erects it.

The lower belly, the inferior coverts of the tail and the thighs are of a brown colour; the tarſus and the toes of a dirty orange; the bill partly black, and partly fleſh-coloured. But the diſtribution of theſe colours ſeems not fixed in that part; for in the ſubjects which we have obſerved, and in thoſe of Aldrovandus, the baſe of the bill was blackiſh, and all the reſt of a fleſh colour; whereas in thoſe examined by Edwards, the point of the bill exhibited the black, which changed by degrees into a dirty orange on the baſe of the bill and on the legs. The under-ſide of the tail ſeemed marbled, the effect produced by the colour of its lower coverts, which are blackiſh and tipt with white.

In the female the head is black like that of the male, but not the neck, nor the quills of the tail and of the wings, which are of a lighter tinge; the colours of the ſcapulary are alſo leſs vivid.

This bird is rather ſmaller than the common Blackbird; its bill, wings, legs, and toes, are proportionably longer. In ſize, figure, and even inſtinct, it is much more analogous to the Ring-Ouzel, for it likewiſe migrates. However, we muſt own, that one of theſe Roſe-coloured Ouzels, which was killed in England, kept company with yellow-billed Blackbirds. Its length, from the point of the bill to the end of the tail, is ſeven inches and three-quarters, and to the extremity of the nails ſeven and a half; its alar extent thirteen or fourteen, and its wings, when cloſed, reached almoſt to the middle of the tail. [A]

[A] Specific character of the *Turdus Roſeus*:—"It is ſomewhat of a carnation colour, its head, its wings, and its "tail, black; the back of its head creſted." It very rarely appears in England. It annually reſorts in great flocks about the river Don, where it breeds among the rocks. The *Turdus Seleucis* of Gmelin is really the ſame ſpecies, which in Syria obtains the name of *locuſt bird*. It viſits Aleppo in the months of July and Auguſt in purſuit of the ſwarms of locuſts; and hence the Turks regard it as ſomewhat ſacred.

Nº 79

THE ROCK SHRIKE

The ROCK BLACKBIRD *.

Le Merle de Roche, Buff.
Turdus Saxatilis, } Gmel.
Lanius Infaustus, } Gmel.
Merula Saxatilis, Ray, Will. and Briss.
The Greater Red-Start, Alb.
The Rock Crow, Penn.

THE name indicates sufficiently the haunts of this bird: it inhabits precipices and mountains; it is found in the wildest parts of Bugey; it sits commonly on the large stones, and constantly without cover; so that it is difficult to get near it with a fowling-piece, for as soon as it perceives the person, it removes to another place. Its shyness seems to be less owing to native wildness, than to its apprehensions of man, and its experience of his artifices. Nor is it so much exposed as many other birds to danger from that quarter. The loss of liberty alone is what it has to dread; for though excellent eating, it is more prized on account of its song, which is soft, varied, and much like that of the Pettychaps. It soon acquires the notes of other birds, and even learns our music. It begins by day-break, and welcomes the return of the morning; and it renews its strain with the setting sun. If during the night we go near its cage with a light, it

* In Italian, *Codirosso Maggiore*, *Corossolo*, *Crosserone*: in German, *Stein-Roetele*, *Stein-Trostel*, *Stein-Reitling*.

immediately ſings; and in the day-time, if it is not warbling, it ſeems humming and preparing new airs.

Theſe birds conceal their neſts with the utmoſt care, and build them in the holes of the rocks, and in the bottom of the moſt inacceſſible caverns. It is with the greateſt difficulty and hazard that we can ſcramble to theſe, which they defend with courage, darting at the eyes of their plunderers.

Each hatch contains three or four eggs. They feed their young with worms and inſects, on which they live themſelves. They can ſubſiſt however on other food, and when they are raiſed in a cage, it ſucceeds well to give them the ſame paſte as the Nightingales. But they muſt be taken from the neſt; for after they have flown, they cannot be enticed into any kind of ſnare; and if they be caught by ſurpriſe, they will never ſurvive their liberty *.

The Rock Blackbirds are found in many parts of Germany, in the Alps, in the mountains of Tyrol, in thoſe of Bugey, &c. I received a female of this ſpecies caught on its eggs the 12th of May; it had built its neſt on a rock in the neighbourhood of Montberd, where theſe birds are very rare and quite unknown; its colours were not ſo bright as thoſe of the male. This laſt is rather ſmaller than the common Blackbird, and entirely different in its proportions.

* FRISCH.

Its wings are very long, ſuch as would ſuit a bird that neſtles in the bottom of caverns; they meaſure thirteen or fourteen inches when expanded, and if cloſed they reach almoſt to the end of the tail, which is only three inches in length. The bill is about an inch.

With regard to the plumage, the head and neck are covered as it were with a cinereous cowl, variegated with ſmall ruſty ſpots. The back is darker near the neck, and lighter near the tail. The ten lateral quills of the tail are ferruginous, and the intermediate brown. The wing-quills and their coverts are of a duſky colour, and edged with a lighter tinge. Laſtly, the breaſt, and all the lower-part of the body, orange, variegated with ſmall ſpeckles, ſome white, others brown; the bill and legs are blackiſh. [A]

[A] There are two kinds of Rock Blackbirds, or Ouzels; a greater and a leſſer. The former has ſometimes been denominated a Crow or Shrike. It is the *Turdus Infauſtus* of Latham, who thus characterizes it: "It is blackiſh, variegated with duſky and "tawny, its head ſpotted with cinereous tawny, its lateral tail-"quills rufous." It is of the ſize of a Thruſh, and occurs chiefly in Italy and the ſouth of Europe.

The Leſſer Rock Ouzel is the one whoſe habits are exhibited in the text. It is the *Turdus Saxatilis* of Latham. Its ſpecific character: "Its head cœrulean, its tail ferruginous."

M

The BLUE OUZEL*.

Le Merle Bleu, Buff.
Turdus Cyanus, Linn. and Gmel.
Merula Cærulea, Briff.
Turdus Solitarius, Klein.
Cyanos, feu *Cærulea Avis*, Ray.
The Indian Mock-Bird, Will.
The Solitary Sparrow, Edw.
The Blue Thrufh, Lath.

THIS bird has the fame ground colour with the Rock Blackbird; that is, a cinereous blue, without any mixture of orange; the fame fize, the fame proportions nearly, the fame tafte for certain kinds of food, the fame fong, the fame habit of fettling on the fummits of mountains, and of building its neft in the moft craggy rocks. In fhort, we might be inclined to refer it to the fame fpecies. Accordingly, many naturalifts have miftaken the one for the other. The colours of its plumage vary fomewhat in the defcriptions, and it is probably fubject to real variations, arifing from the difference of the individuals, that of age, of fex, of climate, &c. The male which Edwards has delineated, Pl. XVIII. was not of an uniform blue throughout; the tinge of the upper-part of the body was deeper

* In Italian, *Merlo Biavo:* in German, *Blau-Vogel, Blau-Stein-Amfel, Klein-Blau-Zimmer* (Little Blue Zimmer).

than

No. 80

THE BLUE THRUSH

than that of the lower; the quills of its tail blackiſh, thoſe of its wings brown, and alſo the great coverts, which are edged with white; its eyes ſurrounded by a yellow circle, the inſide of its mouth orange, its bill and legs of a brown verging on black. There would ſeem to be more uniformity in the plumage of the female.

Belon, who ſaw ſome of theſe birds at Raguſa in Dalmatia, tells us, that they are alſo found in the iſlands of Negropont, Candia, Zante, Corfu, &c.; that they are very much ſought for, on account of their ſong; but he adds, that they do not inhabit France or Italy. However, the arm of the ſea which ſeparates Dalmatia from Italy is no inſurmountable barrier, eſpecially to theſe birds, which, according to Belon himſelf, fly much better than the common Blackbird, and which could at leaſt make the circuit and penetrate into Italy by the State of Venice. Beſides, it is a fact that theſe Ouzels are found in Italy; the one deſcribed by Briſſon, and that figured in our *Pl. Enl.* No. 250, were both ſent from that country. Edwards had learnt from current report that they neſtled on inacceſſible rocks, or old deſerted towers *, and he ſaw ſome which were killed

* M. Lottinger tells me of a Lead-coloured Ouzel which paſſes into the mountains of Lorraine in the months of September and October, which is then much fatter and better taſted than our common Blackbirds, but reſembles neither the male nor the female of that ſpecies. As no deſcription accompanied this note, I cannot decide whether it refers to the Blue Ouzel.

killed near Gibraltar; from which he infers, with great probability, that they are ſpread through the whole of the ſouth of Europe. But this muſt be underſtood of the mountainous tracts, for it is rare to find them in the plains. They commonly lay four or five eggs, and their fleſh, eſpecially when they are young, is reckoned good eating *. [A]

* Belon.

[A] Specific character of the *Turdus Cyanus*:—"Its quills a" "blue-aſh coloured at the margin, its mouth and eye-lids yellow." It is eight inches long, but ſmaller than the Blackbird.

The SOLITARY OUZEL*.

Le Merle Solitaire, Buff.
Turdus Solitarius, Linn. and Gmel.
Merula Solitaria, Briss.
Passer Solitarius, Ray.
The Solitary Thrush, Lath.

THIS also is an inhabitant of the mountains, and famous for its elegant strains. It is well known that Francis I. king of France, took singular pleasure in listening to it; and even at present the male of this species is tamed and sold at a very high price at Geneva and Milan †, and still dearer at Smyrna and Constantinople ‡. The native warble of the Solitary Ouzel is extremely liquid and tender, but rather plaintive, as must be the song of every bird which leads a lonely existence. It remains always single, except in the season of love. At that joyous period, the male and female not only associate together,

* It is probably the Κοσσυφος Βαιος, or the Little Blackbird, of Aristotle, which resembled the Blackbird, only its plumage was brown, its bill not yellow, and it lodged among rocks or on roofs. In modern Greek, Μερολα: in Italian, *Passera Solitaria*; and also *Merulo Solitario*, *Saxatili*, *Stercoroso*, *Merlo Chiappa* (Buttock-Blackbird): in Turkish, *Kajabulbul*, which signifies Rock Nightingale; the Swedish *Sten-Naecktergahl* has the same meaning: in Polish, *Wrobel Osobny*.

† Olina, Gesner, Willughby.

‡ It is sometimes sold in these cities for fifty or a hundred piastres. HASSELQUIST.

gether, but desert in company the wild and dreary heights where they had lived separately, and resort to the milder abodes of man. They seem to seek spectators of their pleasures, and come forward in those intoxicating moments, when other animals court the silence of retreat. But they lodge at a considerable height above the surface, and thus in the midst of population they shun the dangers to which they would be exposed. They build their nest with stalks of grass and feathers in the top of a separate chimney, or on the ruins of an old castle, or on the summit of a large tree, and almost always near a steeple or lofty tower. The male sits whole hours or days upon the vane or weather-cock, and soothes the tedious situation of his mate by a continual warble; but pathetic as are his strains, they are still insufficient to express the warmth and tenderness of his emotions. A solitary bird feels more delicately and ardently than others. Sometimes he rises chanting, flapping his wings, displaying the feathers of his tail, bristling those on his head, and panting with delight, he describes many circles in the air round his beloved mate as the centre.

If the female be scared by any uncommon noise, or by the sight of any new object, she retires into her fort, but soon returns to the nest, which she never abandons.

As soon as the young are hatched, the male ceases to sing, but not to love.; he gives another proof

proof of his affection by sharing in the trouble of rearing the brood, and bringing provisions in his bill. In animals the ardor of love is ever proportioned to the tenderness for the offspring.

They commonly lay five or six eggs; they feed their young with insects, on which, and on grapes, they subsist themselves *. They arrive in April in those countries where they pass the summer, and depart about the end of August; they return every year to the same spot where they first fixed their abode. It is uncommon to see more than two pairs settled in the same tract †.

The young, when they are taken out of the nest, are capable of instruction, and they learn to chant or to prattle. They begin to sing at midnight, on the approach of the light of a candle. When well-treated they can live in a cage eight or ten years. They are found on the mountains in France and Italy ‡, in almost all the islands of the Archipelago, especially in Zira and Nia, where it is said they nestle among the heaps of stones §, and in the island of Corsica, where they are not considered as birds of passage ‖. But in Burgundy, those which ar-

* Willughby, Belon, &c.

† There is every year a pair of them in the belfry of Sainte-Reine, a small town in my neighbourhood, situated on the declivity of a hill of moderate height.

‡ Belon. § Hasselquist.

‖ Artier, Professor of Natural History at Bastia.

rive in the ſpring, and lodge on the chimney tops, and in ruined churches, were never known to ſpend the winter in that province. The Solitary Ouzel may not migrate in Corſica, and yet flit from one part to another, according to the change of ſeaſons, as it does in France.

The ſingular habits of this bird, and the charms of its ſong, have inſpired in the people a ſort of veneration for it. I know ſome places where it is looked upon as lucky, where they would hardly ſuffer its neſt to be diſturbed, and dread its death as a public misfortune.

The Solitary Ouzel is rather ſmaller than the common Blackbird, but its bill is ſtronger and more hooked near the point, and the legs are ſhorter in proportion. Its plumage is brown of different ſhades, and ſpeckled throughout with white, except on the rump, and on the feathers of the wings and tail. Alſo, its neck, throat, breaſt, and the coverts of the wings, are in the male of a blue tinge, with purple reflexions, entirely wanting in the female, which is of an uniform brown, with yellowiſh ſpeckles. In both, the iris is of an orange yellow, the opening of the noſtrils wide, the edges of the bill ſcalloped near the tip, as in almoſt all the Blackbirds and Thruſhes; the inſide of the mouth yellow, the tongue parted into three threads, of which the mid one is the longeſt; twelve quills in the tail, nineteen in each wing, the firſt of which is very ſhort: laſtly, the firſt *phalanx* of the outer toe is is joined to that of the middle one. The total length of the bird is eight or nine inches; its alar extent twelve or thirteen; its tail three; its leg thirteen lines; and its bill fifteen; the wings, when closed, reach beyond the middle of the tail. [A]

[A] Specific character of the *Turdus Solitarius*:—"It is dusky, "a great part of it spotted with white, its tail blackish."

M

FOREIGN BIRDS,

WHICH ARE RELATED TO THE SOLITARY OUZEL.

I.

The PENSIVE THRUSH.

Le Merle Solitaire de Manille, Buff.
Turdus Manillensis, Gmel.

THIS ſpecies ſeems to be intermediate between the Solitary Ouzel and the Rock Blackbird. It has the colours of the latter, and diſtributed partly in the ſame order; but its wings are not ſo long, though when cloſed they reach to two-thirds of the tail. Its plumage is a ſlate-blue, uniform on the head, the hind-part of the neck, and the back; almoſt quite blue on the rump, ſpeckled with yellow on the throat, and on the fore-part of the neck and top of the breaſt. The ſame blue colour is deeper on the coverts of the wings, with ſimilar ſpeckles, though ſcattered more ſparingly, and ſome white ſpots, which are ſtill fewer. The reſt of the under-ſide of the body is orange, ſpeckled with blue and white; the quills of the wings and of the tail are blackiſh, and the latter edged with rufous;

rufous; laſtly, the bill is brown, and the legs almoſt black.

The Penſive Ouzel is nearly of the ſize of the Rock Blackbird; its total length is about eight inches, its alar extent twelve or thirteen, its tail three, its bill only an inch.

The female has no blue or orange in its plumage, but two or three ſhades of brown, which form pretty regular ſpeckles on the head, the back, and all the under-ſide of the body.—Theſe two birds were preſented by M. Sonnerat.

M

II.

The HERMIT THRUSH.

Le Merle Solitaire des Philippines, Buff.
Turdus Eremita, Gmel.

The figure of this bird, its port, and its bill, reſemble thoſe of the Solitary Ouzels, and its plumage is ſomewhat analogous to that of the Penſive Ouzel, but it is rather ſmaller. Each feather in the under-ſide of the body is rufous of various ſhades, and edged with brown. The feathers of the upper-ſide of the body are brown with a double border, the inner blackiſh, and the outer dirty white. The ſmall coverts of the wings have an aſh-caſt, and thoſe of the rump and tail are quite cinereous. The head is olive, verging

verging on yellow, the orbits whitiſh, the quills of the tail and of the wings edged with gray; the bill and legs brown.

The entire length of the Hermit Ouzel is about ſeven inches and a half, its alar extent twelve, and its wings, if cloſed, reach to three-fourths of its tail, which contains twelve quills, and is only two inches and three quarters long.

This bird, which was ſent by M. Poivre, reſembles in ſo many reſpects the Penſive Ouzel, that I ſhould not wonder if it be afterwards found only a variety of age or ſex; eſpecially as it is brought from the ſame country, is ſmaller, and its colour intermediate between thoſe of the male and of the female.

M

FOREIGN BIRDS,

WHICH ARE RELATED TO THE EUROPEAN BLACKBIRDS.

I.

The AFRICAN THRUSH.

Le Jaunoir * *du Cap de Bonne Esperance,* Buff.
Turdus Morio, Linn. and Gmel.
Merula Capitis Bonæ Spei, Briff.

THIS bird has the black and yellow colours of the European Blackbirds: but the black is more brilliant, and has reflexions which in certain positions have a greenish cast. The yellow, or rather the rufous colour, is seen only on the quills of the wings, of which the three first are tipt with brown, and the following with this brilliant black I have mentioned. The same lucid refulgent black occurs on the two middle quills of the tail, and on that part of the middle quills of the wings which is uncovered; all that is hid of these middle quills, and all the lateral quills of the tail, are of a pure black. The bill is of the same black, but the legs are brown.

The African Thrush is larger than the common Blackbird; its length is eleven inches, its

* A word compounded of *Jaune*, yellow, and *Noir*, black; which are the colours of its plumage.

 alar

alar extent fifteen and a half, its tail four; its bill, which is thick and ſtrong, is fifteen lines, and its leg fourteen; its wings, when cloſed, reach not to the middle of its tail.

M

II.

The CRESTED BLACKBIRD of China.

Gracula Criſtatella, Linn. and Gmel.
Merula Sinenſis Criſtata, Briſſ.
Sturnus crinibus cinereis, &c. Klein.
The Chineſe Starling, or *Blackbird*, Edw.
The Creſted Grakle, Lath.

Though this bird is ſomewhat larger than the Blackbird, its bill and legs are ſhorter, and its tail much ſhorter; almoſt all its plumage is blackiſh, with a dull blue tinge, but not gloſſy; a white ſpot appears in the middle of the wings, and impreſſed on the quills, and a little white on the tips of the lateral quills of the tail; the bill and legs are yellow, and the iris of a fine orange. There is a ſmall tuft of pretty long feathers on the forehead, which the bird can briſtle up at pleaſure. But notwithſtanding this mark of diſtinction, and the difference perceived in its proportions, we may perhaps regard it as a variety, produced by climate, of our Yellow-billed Blackbird. It has, like that

that bird, a great facility in learning to whiftle airs, and in repeating words. It is difficult to be brought from China into Europe. Its length is eight inches and a half; its wings, when clofed, reach to the middle of the tail, which is only two inches and a half long, and compofed of twelve quills nearly equal*.

* Specific character of the *Gracula Criftatella*, LINN.—" It " is black, the primary wing-quills white at their bafe, and the " tail-quills at their tips; the bill yellow."

M

III.

The RUFOUS-WINGED THRUSH.

Le Podobé du Senegal, Buff.
Turdus Erythropterus, Gmel.

We are indebted to M. Adanfon for this foreign and new fpecies; its bill is brown, its wings and legs rufous, its wings fhort, its tail long, tapered, marked with white at the extremity of the lateral quills, and of the lower coverts. In every other part the Podobé is of the colour of our Blackbirds, and refembles them in fize, and in the fhape of the bill, which, however, is not yellow.

IV.

The BLACKBIRD of China.

Turdus Perſpicillatus, Gmel.
The Spectacle Thruſh, Lath.

This Blackbird is larger than ours, its legs much ſtronger, its tail longer and differently ſhaped, for it is tapered. The moſt remarkable feature in its plumage, is what appears like a pair of ſpectacles, placed at the baſe of the bill, and extending both ways upon the eyes; the ſides of theſe ſpectacles are nearly of an oval form, and black, ſo that they are diſtinctly defined on the gray plumage of the head and neck. The ſame gray colour, intermixed with a greeniſh tint, is ſpread over the whole of the upperſide of the body, including the wings and the intermediate quills of the tail; the lateral quills are of a much deeper colour; part of the breaſt, and the belly, are of a dirty white, with a little yellow, as far as the lower coverts of the tail, which are rufous. The wings when cloſed extend not far beyond the origin of the tail.

V.

The GLOSSY THRUSH.

Le Vert-Doré, ou *Merle a Longue Queue du Senegal* *, Buff.
Turdus Æneus, Gmel.

The extreme length of this bird, which is about ſeventeen inches, is only two-thirds of that of its tail. Its alar extent by no means correſponds to the ſame proportion, being narrower than that of the common Blackbird, which is a much ſmaller bird. Its bill is alſo proportionably ſhorter, but its legs are longer †. The prevailing colour is the fine gloſſy green that appears in the plumage of Ducks; the only difference is derived from the various tints and reflexions which in different parts it aſſumes. It is blackiſh on the head, with gold colour ſhining through; and on the rump and the two long intermediate quills of the tail are purple reflexions; on the belly and thighs a changing green, with roſe-copper. Almoſt all the reſt of its plumage is of a rich gold green.

* *i. e.* The Golden-Green, or Long-tailed Blackbird of Senegal.

† The meaſures given by Briſſon are theſe:—Total length eighteen inches; from the point of the bill to the end of the nails ten and a half; alar extent fourteen and a quarter; the length of the tail eleven; the bill thirteen lines; the legs eighteen.

There is in the Royal Cabinet a bird exactly like this, only its tail is not near ſo long. It is probably the ſame bird, but caught in the time of moulting *.

* It is titled, *The Green Ouzel of Senegal.*

VI.

The CRESCENT BLACKBIRD of America.

Le Fer-a-Cheval, ou *Merle a Collier d'Amerique,* Buff.
Alauda Magna, Linn. and Gmel.
Sturnus Ludovicianus, var. Lath.
Merula Americana Torquata, Briſſ.
The Large Lark, Cateſby.
The Creſcent Stare, Penn. and Lath.

The only black part of the plumage of this bird is a mark ſhaped like a horſe-ſhoe, which deſcends upon the breaſt, and a bar of the ſame colour riſing on each ſide under the eye, and extending backwards. The firſt of theſe ſpots ſeems, from its determined figure, to be the moſt characteriſtic of this ſpecies, and diſtinguiſhes it the beſt from the other collared Blackbirds. This horſe-ſhoe is traced on a yellow ground, which is the colour of the throat and of all the under-ſide of the body, and which appears again between the bill and the eyes; brown predominates on the head and behind the neck, and light gray on the ſides. Alſo the top of the head

head is marked with a whitiſh ray; all the upper-ſide of the body is of a partridge-gray; the quills of the wings and of the tail * are brown, with ſome ruſty ſpots; the legs brown and very long; and the bill, which is almoſt black, is ſhaped like that of our Blackbirds; like them alſo it ſings agreeably in the ſpring, though it has not the ſame extent of notes. It ſcarcely eats any thing but the ſmall ſeeds which it finds on the ground †; in which reſpect it reſembles the Larks, though it is much larger, exceeding even our Blackbird, nor is its hind-nail lengthened as in the Larks. It perches on the top of buſhes, and its tail is obſerved to have a briſk motion upwards and downwards. In fact, it is neither a Lark nor a Blackbird; and yet of all the European birds, it reſembles the latter the moſt. It is found not only in Virginia and Carolina, but in almoſt the whole continent of America ‡.

The ſubject examined by Cateſby weighed three ounces and a quarter; its extreme length ten inches, its bill fifteen lines, its legs eighteen; its wings when cloſed reached to the middle of its tail. [A]

* Linnæus ſays, that the three lateral quills of the tail are partly white. *Syſt. Nat.* Edit. x. p. 167.

† For inſtance, thoſe of the Yellow-flowered *Ornithogalum.*

‡ Linnæus aſſerts that it occurs alſo in Africa.

[A] Authors are much divided with reſpect to the claſſification of this bird: Lawſon and Cateſby call it a Lark; Briſſon reckons it a Blackbird; Pennant makes it a Stare; and Latham regards it

as

as a variety of the Louiſiana Stare already deſcribed; and Gmelin ſeems inclined to the ſame opinion. It is thus characterized by Mr. Latham:—" Above it is variegated with ruſty brown and " blackiſh, below yellow, with a black curved ſtripe on the breaſt, " the three lateral quills of the tail white," It lives in ſavannas, and is eſteemed good eating. In the State of New-York it appears in the beginning of April, breeds in June, and retires in September or October. It neſtles on the ground, and its eggs are whitiſh.

M

VII.

The GREEN BLACKBIRD of Angola.

The Blue and Green Daw, Edw.

The upper-part of the body, the head, the neck, the tail, and the wings, are of an olive green; but brown ſpots appear on the wings, and the rump is blue. On the back, and on the fore-ſide of the neck, is a mixture of blue with green; the blue again occurs on the upper-part of the throat: violet predominates on the breaſt, the belly, the thighs, and the feathers which cover the ears: laſtly, the lower coverts of the tail are of an olive yellow, the bill and legs of a deep black.

This bird is of the ſame ſize with the fifty-third Thruſh of Briſſon; the proportions are likewiſe the ſame, but the plumage of the latter is different, being entirely of a fine duck-green, with

with a ſpot of ſteel-violet on the anterior part of the wing.

Theſe birds are nearly of the bulk of our Blackbird, their length being nine inches, their alar extent twelve and a quarter, their bill eleven or twelve lines; their wings when cloſed reach to the middle of the tail, which conſiſt of twelve equal quills.

It is probable that theſe two birds belong to the ſame ſpecies, but I cannot decide which is the original ſtem, and which the collateral branch. [A]

[A] This bird is a variety of the *Shining Thruſh* of Latham, the *Turdus Nitens* of Linnæus, and the *Turdus Viridis Angolenſis* of Briſſon. The character of the ſpecies is, "That it is green, with "a ſpot of ſhining violet on the coverts of the wings."

M

VIII.

The GILDED THRUSH.

Le Merle Violet du Royaume de Juida, Buff.

The plumage of this bird is painted with the ſame colours as the preceding, that is, with violet, green, and blue, but differently diſtributed; violet is ſpread without any mixture on the head, the neck, and all the under-part of the body; blue on the tail and its upper coverts; and laſtly, ly, green on the wings; but these have besides a blue stripe near their inner margin.

This bird is also of the same size with the preceding: it appears to have the same port; and as it comes from the same climates, I should be tempted to refer it to the same species, were it not longer winged, which implies other instincts and habits. But as the length of the wings in dried birds depends greatly on the mode of preparing them, we cannot admit the circumstance just mentioned to constitute a specific difference; and it will be prudent to wait the decision of accurate observation.

IX.

The CEYLON THRUSH.

Le Plastron-Noir de Ceilan, Buff.
Turdus Zeylonus, Linn. and Gmel.
Merula Torquata Capitis Bonæ Spei, Briss.
The Green Pye of Ceylon, Edw.

I bestow a separate name on this bird, because those who have seen it do not agree with regard to the species to which it belongs. Brisson makes it a Blackbird, and Edwards a Pie or a Shrike. For my own part, I conceive it to be a Ring-Ouzel, not venturing, however, to decide, till farther information clear up the subject.

ject. It is ſmaller than the Blackbird, and its bill proportionably ſtronger. Its total length is ſeven inches and a half, its alar extent eleven, its tail three and a half, its bill twelve or thirteen lines, and its legs fourteen; its wings when cloſed reach beyond the middle of its tail, which is ſomewhat tapered.

The black breaſt-piece which diſtinguiſhes this bird is the more conſpicuous, as it is bounded above and below by a lighter colour, for the throat and all the under-part of the body is of a pretty bright yellow. From the two ends of the upper-margin of this breaſt-piece ariſe two cords of the ſame colour, which firſt aſcending on each ſide towards the head, define the beautiful yellow orange plate on the throat, and then bending under the eyes, terminate at the baſe of the bill, where they are in a manner inſerted. Two yellow eye-brows, which take their origin cloſe to the noſtrils, embrace the eyes above, and form a contraſt to the black cords. All the upper-part of this bird is olive; but that colour ſeems to be tarniſhed by a mixture of cinereous on the top of the head, and on the contrary to brighten on the rump, and on the outer edge of the wing-quills; the largeſt of theſe are tipt with brown; the two intermediate ones of the tail are of an olive green, and alſo the whole of the under-part of the body; and the ten lateral ones are black, tipt with yellow.

The

The female has neither the black breaſt-piece nor the black cords. Its throat is gray; its breaſt and belly of a greeniſh yellow, and all the upper-ſide of the body of the ſame colour, but deeper. In general, the female differs little from the bird figured *Pl. Enl.* No. 358, under the name of the *Orange-bellied Blackbird of Senegal.*

Briſſon has ſuppoſed that this bird is a native of the Cape of Good Hope; and indeed it was brought from that place by the Abbé de la Caille. But if we believe Edwards, it belongs to a more diſtant climate, that of the iſland of Ceylon. That naturaliſt obtained accurate information on this ſubject from John Gideon Loten, who had been governor of Ceylon, and who, on his return from India, preſented ſeveral birds of that country to the Royal Society, and among the reſt a Ceylon Thruſh. Edwards introduces here an obſervation which we have already anticipated, but which it may not be improper to repeat. The Cape of Good Hope is the general rendezvous of ſhips trading to the Eaſt, and it may often happen, that in touching there, birds may be left which afterwards are miſtaken for natives of the extremity of Africa.

X. The

X.

The ORANGE-GREEN or the ORANGE-BELLIED BLACKBIRD of Senegal.

Turdus Chrysogaster, Gmel.
The Orange-bellied Thrush, Lath.

The principal colours of this new ſpecies are green and orange; a fine deep green, with reflexions which are variouſly ſhaded with yellow, is ſpread over the whole of the upper-part of the body, including the tail, the wings, the head, and even the throat, but is not ſo deep on the tail. The under-part of the body, from the throat downwards, is of a ſhining orange. When the wings are cloſed, there appears a train of white which belongs to the outer edge of ſome of the quills. The bill is brown, and alſo the legs.—This bird is ſmaller than the Blackbird; its length is about eight inches; its alar extent eleven and a half; its tail two and three-quarters, and its bill eleven or twelve lines.

VARIETY *of this* BIRD.

The preceding bird reſembles much the female of the Ring-Ouzel of Ceylon; but it is equally related to the *Blackbird of the Cape of Good Hope*, No. 221, which I call *Orange-Blue (oranbleu)*; for the whole of the under-part of its body is orange, from the throat to the lower belly inclusive; and blue is spread over the upper-part from the base of the bill to the end of the tail. This blue consists of two shades, the deeper of which edges each feather, whence results an agreeable and regular variety. The bill and legs are black, and also the quills of the wings; but many of the middle ones have a white-gray margin: lastly, the tail-quills are the most uniform in regard to colour.

XI.

The BROWN BLACKBIRD of the CAPE of GOOD HOPE.

Turdus Bicolor, Gmel.
The White-rumped Thrush, Lath.

We are indebted to Sonnerat for this new species. It is nearly the size of the Blackbird; its total length ten inches, and its wings extend a little beyond the middle of the tail. Almost all its plumage is of a varying brown, with reflexions of dusky green: the belly and rump are white.

XII. The

XII.

The BANIAHBOU of Bengal.

Turdus Canorus Lanius Fauſtas, Linn. and Gmel.
Merula Bengalenſis, Briſſ.
The Brown Indian Thruſh, Edw.
The Crying Thruſh, Lath.

Its plumage is every where brown; deeper on the upper-part of the body, lighter on the under, and alſo on the edge of the coverts and wing-quills; the bill and legs are yellow; the tail tapered, about three inches long, and extending half its length beyond the cloſed wings. Such are the chief circumſtances which characterize this foreign bird, the ſize of which ſomewhat exceeds that of the Throſtle.

Linnæus informs us, on the authority of ſome Swediſh naturaliſts who had travelled into Aſia, that the ſame bird occurs in China; but it ſeems there to have been affected by the climate, being gray above and ruſt-coloured below, with a white ſtreak on each ſide of the head. The epithet of *canorus*, which Linnæus beſtows on it, no doubt from accurate information, implies that theſe foreign Blackbirds have an agreeable warble.

XIII.

The CINEREOUS BLACKBIRD.

L'Ourovang, ou *Merle Cendré de Madagascar*, Buff.
Turdus-Urovang, Gmel.
Merula Madagascariensis Cinerea, Briss.

The name Cinereous Blackbird gives a very just idea of the predominant colour of the plumage: but the intensity is not every where the same: it is very deep, almost black, with a slight tinge of green on the long and narrow feathers that cover the head: it is lighter without mixture of other tinge on the quills of the tail and of the wings, and on the great coverts of the latter. It has an olive cast on the upper-part of the body, on the small coverts of the wings, on the neck, on the throat, and on the breast. Lastly, it is lighter under the body, and about the lower belly, and there is a slight tinge of yellow.

This Blackbird is nearly as large as our Red-Wing, but its tail is rather longer, its wings somewhat shorter, and its legs much shorter. Its bill is yellow, as in our Blackbirds, marked near the end with a brown ray, and furnished with some bristles round the base; its tail consists of twelve equal quills, and its legs are of a brown colour.

XIV. The

XIV.

The PIGEON THRUSH.

Le Merle des Colombiers, Buff.
Turdus Columbinus, Gmel.

This bird is called, in the Philippines, the *Pigeon-house Stare*; becaufe it is naturally familiar, and feeks the conveniencies which the dwellings of men afford, and neftles even in the pigeon-houfes. But it refembles the Blackbird more than the Stare, in the fhape of its bill and legs, and in the proportions of its wings, which only reach the middle of the tail, &c. Its bulk is nearly that of the Red-Wing, and its plumage confifts of one colour, though not uniform; this is a varying green, which, according to its pofition, has different fhades and reflexions. This fpecies is new, and we are indebted for it to Sonnerat. There are alfo found in the collection which he brought from the Cape of Good Hope, fome individuals that evidently belong to the fame fpecies, but which differ in having their rump white, both on the upper and under furface, and in being fmaller. Muft this be afcribed to climate or to age * ?

* Specific character of the *Turdus Columbinus*:—" It is green, " with different coloured reflexions."

M

XV.

The OLIVE THRUSH.

Le Merle Olive du Cape de Bonne Esperance, Buff.
Turdus Olivaceus, Linn. and Gmel.

The upper-part of the body of this bird, including what appears of the quills, of the tail, and of the wings, when they are closed, is of an olive-brown; the neck and the breast are of the same colour as the throat, but without streaks; all the rest of the under-part of the body is of a fine fulvous colour: lastly, the bill is brown, as well as the legs, and the inside of the quills of the wings and the lateral quills of the tail.

This Blackbird is as large as a Red-Wing; its alar extent near thirteen inches, and its total length eight and a quarter; the bill is ten lines, the leg fourteen; the tail, which consists of twelve equal quills, is three inches long; and the wings, when closed, reach only half its length *.

* Specific character of the *Turdus Olivaceus*:—" It is somewhat " dusky, below dusky."

M

XVI. The

XVI.

The BLACK-THROATED THRUSH.

Le Merle à Gorge Noire de Saint Domingue, Buff.
Turdus Ater, Gmel.

The black on the throat of this bird extends on the one hand below the eye, and even on the ſpace between the eye and the bill; and on the other it deſcends upon the neck as far as the breaſt. It is beſides edged with a broad rufous border, with different ſhades of brown, which extends upon the eyes and upon the fore-part of the top of the head; the reſt of the head, the poſterior ſurface of the neck, the back, and the ſmall coverts of the wings, are grayiſh brown, variegated ſlightly with ſome browner tints. The great coverts of the wings, as well as the quills, are of a blackiſh brown, edged with light gray, and ſeparated from the ſmall coverts by an olive yellow line belonging to theſe ſmall coverts. The ſame olive-yellow predominates on the rump, and on all the under-part of the body; but under the body it is variegated with ſome black ſpots, which are pretty broad, and ſcattered thinly over the whole ſpace between the black piece of the throat and the legs. The tail is of the ſame gray as the upper-part of the body, but in its middle only; the lateral quills being edged on the outſide with a blackiſh colour; the bill and the legs are black.

This bird, which has not been hitherto deſcribed, is nearly of the bulk of the Red-Wing; its total length is about ſeven inches and a half, its bill one inch, its tail three; and its wings, which are very ſhort, reach ſcarcely the fourth of its tail.

M

XVII.

The CANADA BLACKBIRD.

This reſembles the moſt the Mountain Blackbird, which is only a variety of the Ring Ouzel. It is ſmaller, but its wings bear the ſame proportion to its tail, not reaching beyond the middle, and the colours of its plumage, which are not very different, are diſtributed in the ſame manner. The ground colour is conſtantly dark-brown, variegated with lighter ſhades in every part, except in the quills of the tail and of the wings, which are of an uniform blackiſh brown. The coverts of the wings have reflexions of a deep but ſhining green; all the other feathers are blackiſh, and terminated with rufous, which, disjoining them from one another, produces a regular variety, ſo that the feathers may be counted from the rufous ſpots.

M

XVIII. The

XVIII.

The INDIAN OLIVE BLACKBIRD.

Turdus Indicus, Gmel.
Merula Olivacea Indica, Briſſ.
The Indian Thruſh, Lath.

All the upper-part of this bird, including the quills of the tail, and thoſe uncovered of the quills of the wing, are of a deep olive-green. All the under-part is of the ſame ground-colour, but of a lighter tinge, and bordering upon yellow. The inner webs of the wing-quills are brown, edged partly with yellow; the bill and legs are almoſt black.—This bird is larger than the Red-Wing; its whole length is eight inches, its alar extent twelve and a half, its tail three and a half, its bill thirteen lines, its leg nine, and its wings when cloſed reach to the middle of its tail.

M

XIX.

The INDIAN CINEREOUS BLACKBIRD.

Turdus Cinereus, Gmel.
Merula Cinerea Indica, Briſſ.
The Aſh-coloured Thruſh, Lath.

The colour of the upper-part of the body is deeper than that of the under. The great coverts and the quills of the wings are edged with white-gray on the outſide; but the middle quills have this edging broader. They have likewiſe another border of the ſame colour on the inſide, from their origin, to two-thirds of their length. Of the twelve tail-quills, the two middle ones are cinereous, like the upper-part of the body; the two following are partly of the ſame colour, but their inſide is black: the eight others are entirely black, as alſo the bill, the legs, and the nails. The bill has ſome blackiſh briſtles near the angles of its opening.—This bird is ſmaller than the Red-Wing; it is ſeven and a quarter in length, twelve and two-thirds alar extent; its tail is three inches, its bill eleven lines, and its leg ten.

XX.

The BROWN BLACKBIRD of Senegal.

Turdus Senegalenſis, Gmel.
Merula Senegalenſis, Briſſ.
The Senegal Thruſh, Lath.

Nothing can be more uniform and ordinary than the plumage of this bird, or more eaſy to deſcribe. It is grayiſh brown on the upper and anterior parts, dirty white on the under-part, brown on the quills of the tail and of the wings, and alſo on the bill and legs. It is not ſo large

as

as the Red-Wing, but its tail is longer, and its bill ſhorter. Its whole length, according to Briſſon, is eight inches; its alar extent eleven and a half, its tail three and a half, its bill nine lines, its leg eleven. Its wings do not reach farther than the middle of its tail, which conſiſts of twelve quills.

XXI.

The TANOMBE, or the MADAGASCAR BLACKBIRD.

Turdus Madagaſcarienſis, Gmel.
Merula Madagaſcarienſis, Briſſ.
The Madagaſcar Thruſh, Lath.

I have retained the name by which this bird is known in its native region. It is to be wiſhed that travellers would thus preſerve the real names of the foreign birds; we ſhould then be able to diſtinguiſh the ſpecies to which each obſervation applied.

The *Tanombé* is rather ſmaller than the Red-Wing; its plumage is in general of a very deep brown on the head, neck, and all the upper-part of the body; but the coverts of the tail and wings have a tinge of green. The tail is a gold green, edged with white, as alſo the wings, which have, beſides, ſome violet changing into green

green at the tips of the great quills, a colour of polifhed fteel on the middle quills and the great coverts, and an oblong mark of fine gold-yellow on the fame middle quills. The breaft is of a rufous brown, the reft of the under-part of the body white; the bill and legs are black, and the tarfus very fhort. The tail is fomewhat forked; the wings reach only to the middle, but its alar extent is greater in proportion than in the Red-Wing. I may obferve, that in a fubject which I had occafion to fee, the bill was more hooked at the point than reprefented in the figure, and in this refpect the Tanombé feems to refemble the Solitary Blackbird.

XXII.

The MINDANAO BLACKBIRD.

Turdus Mindanenfis, Gmel.
The Mindanao Thrufh, Lath.

The fteel colour which appears on part of the wings of the Tanombé, is, in the Mindanao Blackbird, fpread over the head, the throat, the neck, the breaft, and all the upper-part of the body as far as the end of the tail. The wings have a white bar near their outer margin, and the reft of the under-part of the body is white.

This

This bird exceeds not ſeven inches in length, and its wings reach only the middle of the tail, which is ſomewhat tapered.—It is a new ſpecies, introduced by Sonnerat.

Daubenton the younger has obſerved another individual of the ſame kind, in which the ends of the long quills of the wings and of the tail are of a deep varying green, with ſeveral ſpots of wavy-violet on the body, but chiefly behind the head. It is perhaps a female, or elſe a young male.

XXIII.

The GREEN BLACKBIRD of the Iſle of France.

Turdus Mauritianus, Gmel.
The Mauritius Thruſh, Lath.

The plumage of this bird is quite uniform, all the outſide being bluiſh green, verging to brown, but its bill and legs cinereous. It is ſmaller than the Red-Wing: its length is about ſeven inches, its alar extent ten and a half, its bill ten lines, and its wings reach to the third of its tail, which is only two inches and a half. The feathers that cover the head and neck are long and narrow.—It is a new ſpecies.

XXIV. The

XXIV.

The BLACK CASQUE, or the BLACK-HEADED BLACKBIRD of the Cape of Good Hope.

Though at firſt ſight this bird ſeems to reſemble moſt in its plumage that of the following article, the *Brunet*, and particularly the *Yellow-rumped Blackbird of Senegal*, which I conſider as a variety of the ſame ſpecies, we ſtill perceive obvious differences in its colour, and more important ones in the proportions of its limbs. It is not ſo large as the Red-Wing; its total length nine inches, its alar extent nine and a half, its tail three and two-thirds, its bill thirteen lines, and its leg fourteen. Its wings, therefore, ſpread leſs than thoſe of the Brunet, but its bill, tail, and legs, are proportionably longer. Its tail is alſo of a different form, and conſiſts of twelve tapered quills; each wing has nineteen, of which the longeſt are the fifth and the ſixth.

With regard to its plumage, it reſembles that bird in the brown colour of the upper-part of its body, but it differs by the colour of its helmet, which is of a ſhining black; by the rufous colour of its rump, and of the upper-coverts of its tail; by the ruſty caſt of its throat, and of the whole of the under-part of its body, as far as the

the lower coverts of the tail inclufively; by the fmall brown ray on the flanks; by the fmall white fpot which appears on the wings, and which belongs to the large quills; by the blackifh colour of the quills of the tail; and laftly, by the white mark which terminates the lateral ones, and which is larger as the quill is nearer the outfide.

XXV.

The BRUNET of the Cape of Good Hope.

Turdus Capenfis, Linn. and Gmel.
Merula Fufca Capitis Bonæ Spei, Briff.
The Brunet Thrufh, Lath.

The predominant colour of the plumage of this bird is deep brown, which is fpread over the head, the neck, all the upper-part of the body, the tail, and wings; it is rather lighter on the breaft and fides, has a yellowifh caft on the belly and thighs, and gives place to a beautiful yellow on the lower coverts of the tail. This yellow fpot is the more confpicuous, as it is contrafted with the colour of the quills of the tail, which are of a ftill deeper brown below than above. The bill and legs are entirely black.

This

This bird is not larger than a Lark; its wings meaſure ten inches and a half acroſs, and hardly reach to the third of its tail, which is near three inches long, and conſiſts of twelve equal quills *.

* Specific character of the *Turdus Capenſis*:—"It is duſky, its "belly ſomewhat yellowiſh, its vent yellow."

VARIETY *of the* CAPE BRUNET.

The bird repreſented *Pl. Enl.* No. 317, by the name of the *Yellow-rumped Blackbird of Senegal* *, is much analogous to the Brunet, only it is rather larger, and its head and throat are black. The remaining parts are of the ſame colour in both, and nearly of the ſame proportions; which would lead us to ſuppoſe that it is a variety produced by difference of age or of ſex. But having occaſion afterwards to obſerve that, among a great number of birds ſent by Sonnerat, many marked "Cape Blackbirds" were exactly like the ſubject deſcribed by Briſſon, and not one with a black head and throat, it ſeems more probable that the bird, No. 317, is only a variety derived from climate. The bill of this bird is broader at the baſe, and more curved than that of the ordinary Blackbird.

* *Merle a cul-jaune du Senegal.*

XXVI. The

XXVI.

The BROWN JAMAICA BLACKBIRD.

Turdus Aurantius, Gmel.
Merula Jamaicensis, Briss.
Merula Fusca, Ray, Sloane, and Klein.
The White-chinned Thrush, Lath.

Deep brown is the predominant colour of the head, the upper-part of the body, the wings, and the tail; brown of a lighter shade on the fore-side of the breast and of the neck, dirty white under the belly, and on the rest of the lower-part of the body. The most remarkable feature in this bird is, that the throat and bill are white, and the legs orange. Its extreme length is six inches four lines, its alar extent nine inches and some lines, its tail two inches and eight or nine lines, its leg two inches and a quarter, its bill eleven lines; all English measure. It appears then that it is not so large as our Red-Wing. It generally haunts the mountains and forests, and is esteemed good eating. All that Sloane informs us, with respect to the interior structure of this bird, is, that its fat is of an orange-yellow *.

* Specific character of the *Turdus Aurantius*:—" It is blackish-brown, the upper-part of its throat and its belly whitish, its bill and legs orange."

XXVII. The

XXVII.

The CRAVATED BLACKBIRD of Cayenne.

Turdus Cinnamomeus, Gmel.
The Black-breaſted Thruſh, Lath.

The cravat of this Blackbird is very broad, of a fine black edged with white; it extends from the baſe of the lower mandible, and even from the ſpace included between the upper mandible and the eye, as far as the middle of the breaſt, where the white border widens, and is marked with tranſverſe rays of black: it covers the ſides of the head as far as the eyes, and incloſes three-fourths of the circumference of the neck. The coverts of the wings are of the ſame black as the collar; but the ſmall ones are tipt with white, which produces ſpeckles of that colour; and the two rows of great coverts have a fulvous edging. The reſt of the plumage is cinnamon colour, but the bill and legs are black.

This Blackbird is ſmaller than our Red-Wing; the point of its bill is hooked as in the Solitary Thruſhes. Its whole length is about ſeven inches, its tail two and a half, its bill eleven lines, and its wings, which are ſhort, extend but a little way beyond the origin of the tail.

M

XXVIII. The

XXVIII.

The CRESTED BLACKBIRD of the Cape of Good Hope.

Turdus Cafer, Linn. and Gmel.
The Cape Thrush, Lath.

The creſt is not permanent; it conſiſts of long narrow feathers, which naturally recline on the top of the head, but which the bird can briſtle at pleaſure. Its colour, and that of the head and the breaſt, is a fine black, with violet reflexions; the fore-ſide of the neck and breaſt have the ſame wavy gloſs on a brown ground. This brown is ſpread on all the upper-part of the body, and extends over the neck, the coverts of the wings, part of the tail-quills, and even under the body, where it forms a ſort of broad cincture which paſſes under the belly; but in all theſe places it is ſoftened by a whitiſh colour which edges and defines each feather, in the ſame way nearly as in the Ring Ouzel.

The lower coverts of the tail are red, the upper white, the abdomen alſo white, and the bill and legs black. The corners where the bill opens are ſhaded with long black briſtles projecting forwards. This Blackbird is ſcarcely larger than the Creſted Lark. Its wings meaſure eleven or twelve inches acroſs, and when cloſed do not reach the middle of the tail. The longeſt fea thers are the fourth and fifth, and the firſt is the ſhorteſt of all *.

* Specific character of the *Turdus Cafer*:—" It is ſomewhat " creſted, its rump and belly white, its vent red."

XXIX.

The AMBOYNA BLACKBIRD.

Turdus Amboinenſis, Gmel.
The Amboina Thruſh, Lath.

I allow this bird to remain in the place aſſigned it by Briſſon, though I am not quite certain whether it really belongs to this genus. Seba, who firſt noticed it, tells us that he ranged it among the Nightingales, on account of the ſweetneſs of its ſong; it not only chants its loves in the ſpring, but erects its long beautiful tail, and bends it in a ſingular manner over its back. All the upper-part of its body is reddiſh brown, including the tail and the wings, except that theſe are marked with a yellow ſpot; all the under-part of the body is of this laſt colour, but the lower ſurface of the tail-quills is golden. Theſe are twelve in number, and regularly diminiſhing.

XXX.

The BLACKBIRD of the Isle of Bourbon.

Turdus Borbonica, Gmel.
The Bourbon Thrush, Lath.

The size of this bird is nearly that of the Crested Lark; it is seven inches and a half long, and eleven and one-third across the wings; its bill ten or eleven lines, its legs the same, and its wings reach not to the middle of its tail, which is three inches and a half long, and consequently almost half the whole length of the bird.

The top of the head is covered with a sort of black cap; all the rest of the upper-part of the body, the small coverts of the wings, the whole of the tail and breast, are of an olive-ash colour; the rest of the under-part of the body is olive, verging on yellow, except the middle of the belly, which is whitish. The great coverts of the wings are brown, with some mixture of rufous; the wing-quills are parted by these two colours; the brown being placed within and beneath, the rufous before. We must, however, except the three middle-quills, which are entirely brown; those of the tail are brown also, and intersected near their end by two bars of different shades of brown, but from the sameness of the ground colour they are very indistinct: the bill and legs are yellowish.

 XXXI. The

XXXI.

The DOMINICAN BLACKBIRD of the Philippines.

Turdus Dominicanus, Gmel.
The Dominican Thrush, Lath.

The length of the wings is one of the moſt remarkable characters of this new ſpecies; they reach as far as the tail. Their colour, as well as that of the under-ſide of the body, is brown, on which appear a few irregular ſpots of the colour of poliſhed ſteel, or rather of changing violet *. This brown ground aſſumes a violet caſt at the origin of the tail, and a greeniſh at the end; it is lighter on the ſide of the neck, and becomes whitiſh on the head and all the lower-part of the body. The bill and legs are light brown.

This bird is ſcarcely ſix inches long. It is a new ſpecies, for which we are indebted to Sonnerat.

* Theſe violet ſpots, irregularly ſcattered on the upper-ſide of the body, have led Daubenton the younger to ſuppoſe, that this individual was killed at the cloſe of the moulting ſeaſon, before the true colours of its plumage had ſtability.

XXXII.

The GREEN CAROLINA BLACKBIRD.

Cateſby, who obſerved this bird in its native region, informs us, that it is ſcarcely larger than a Lark,

Lark, and its figure is nearly the ſame; that it is extremely ſhy, and conceals itſelf dexterouſly; that it haunts the banks of the large rivers, two or three hundred miles from the ſea, and flies with its feet extended backwards (as uſual in thoſe of our own birds, whoſe tails are very ſhort), and that its ſong is loud. It probably ſubſiſts on the ſeeds of the purple-flowered night-ſhade.

All the upper-part of the body is of a dull green, the eye is almoſt encircled with white, the lower jaw delicately edged with the ſame colour; the tail brown; the under-ſides of the body, except the lower belly, which is whitiſh, the bill and the legs, black: the quills of the wings do not reach much beyond the origin of the tail.

The total length of the bird is about ſeven inches and a quarter, its tail three, its leg twelve lines, its bill ten.

XXXIII.

The TERAT BOULAN, or the INDIAN BLACKBIRD.

Turdus Orientalis, Gmel.
Merula Indica, Briſſ.
The Aſh-rumped Thruſh, Lath.

The characters of this ſpecies are theſe: The bill, legs, and toes, are proportionably ſhorter than in the others; the tail is tapered, but differently from ordinary; the ſix middle quills are of equal lengths, and it is properly the three lateral quills on each ſide that are tapered. The upper-part of the body, the neck, the head, and the tail, are black, the rump cinereous, and the three lateral feathers on each ſide tipt with white. The ſame white prevails on all the under-part of the body and of the tail, on the fore-ſide of the neck, and of the throat, and extends both ways over the eyes; but on each ſide a ſmall black ſtreak riſes at the baſe of the bill, and ſeems to paſs under the eye, and extend beyond it. The great quills of the wings are blackiſh, edged half-way with white on the inſide; the middle quills, and alſo the great coverts, are likewiſe edged with white, but on the outſide, and through their whole length.

This bird is rather larger than the Lark; its alar extent is ten inches and a half, and its wings extend a little beyond the middle of its tail. Its length, from the point of the bill to the end of the tail, is ſix inches and a half, and to the end of the nails five and a half; the tail is two and a half, the bill eight lines and a half, the leg nine, and the middle toe ſeven.

XXXIV. The

XXXIV.

The SAUI JALA, or the GOLDEN BLACKBIRD of Madagafcar.

Turdus Nigerrimus, Gmel.
Merula Madagafcarienfis Aurea, Briff.
The Black-cheeked Thrufh, Lath.

This fpecies, which is an inhabitant of the ancient continent, retains in part the plumage of our Blackbirds. Its bill, legs, and nails, are blackifh; it has a fort of collar of fine velvet black, which paffes under the throat, and extends only a little beyond the eyes; the quills of the tail and of the wings, and the plumage of the reft of the body, are always black, but edged with lemon colour, as they are edged with gray in the Ring Ouzel; fo that the fhape of each feather is beautifully defined on the contiguous feathers which it covers.

This bird is nearly of the fize of the Lark; its alar extent is nine inches and a half, and its tail is fhorter than in our Blackbirds, in proportion to the total length of the bird, which is five inches and three quarters, and alfo in proportion to the length of its wings, which ftretch almoft to two-thirds of its tail. The bill is ten lines, the tail fixteen, the legs eleven, and the middle toe ten.

XXXV.

The SURINAM BLACKBIRD.

Turdus Surinamus, Gmel.
Merula Surinamensis, Briss.
The Surinam Thrush, Lath.

We find in this American Blackbird the same ground colour that predominates in the common Blackbird. It is almost entirely of a shining black, but diversified by other tints. On the crown of the head is a yellowish fulvous plate; on the breast are two marks of the same colour, but of a lighter shade; on the rump is a spot of the same hue; on the wings is a white line that borders them from their origin to the third joint; and lastly, under the wings is white, which prevails over all the lower coverts: so that in flying this bird discovers as much white as black. Its legs also are brown, and its bill only blackish; and also the wing-quills, and all those of the tail, except the two first and the last, which are a yellowish fulvous colour at their origin, but only in the inside.

The Surinam Blackbird is not larger than a Lark; its whole length is six inches and a half; its alar extent nine and a half; its tail three nearly; its bill eight lines, and its legs seven or eight; lastly, its wings stretch beyond the middle of its tail.

XXXVI. The

XXXVI.

The PALMISTE.

Turdus Palmarum, Linn. and Gmel.
The Palm Thrush, Lath.

This bird owes its name to its habits of lodging and neftling in palm-trees, where it no doubt gathers its food. Its bulk is equal to that of the Lark; its length is fix inches and a half, its alar extent ten and one-third, its tail two and a half, and its bill ten lines.

What ftrikes us firft in the plumage is a fort of large black cap, which defcends both ways lower than the ears, and is marked on each fide with three white fpots, the one near the forehead, the other above the eye, and the third below it. The neck is afh-coloured behind where it is not covered by this black cap, and white before, as alfo the throat. The breaft is cinereous, and the reft of the under-part of the body is white-gray. The upper-part of the body, including the fmall coverts of the wings, and the twelve quills of the tail, is of a beautiful olive-green; the parts of the wing-quills that appear are of the fame colour, and the reft is brown. The bill and legs are cinereous.—The wings ftretch a little beyond the middle of the tail.

The bird, which Briffon has made another fpecies of the Palmifte, differs from the preceding in in nothing but that its cap, inſtead of being entirely black, has an aſh-coloured bar on the top of the head, and has rather leſs white under the body. But in every other reſpect the reſemblance is exact, and not a word of the deſcription needs to be altered: and as they inhabit the ſame country, I do not heſitate to conclude that theſe two individuals belong to the ſame ſpecies, and I am diſpoſed to think that the firſt is the male, and the ſecond the female.

XXXVII.

The WHITE-BELLIED VIOLET BLACK-BIRD of Juida.

Turdus Leucogaſter, Gmel.
The Whidah Thruſh, Lath.

The name of this bird contains almoſt a complete deſcription of its plumage. I need only add, that the great wing-quills are blackiſh, that the bill is of the ſame colour, and the legs cinereous. It is rather ſmaller than the Lark; its length is about ſix inches and a half, its alar extent ten and a half, its tail ſixteen lines, its bill eight, its legs nine.—The wings ſtretch three-fourths of the tail.

XXXVIII. The

XXXVIII.

The RUFOUS BLACKBIRD of Cayenne.

Turdus Rufifrons, Gmel.
The Rufous Thrush, Lath.

The front and fides of its head, the throat, and all the fore-part of its neck and belly, are rufous. The top of the head, and all the upper-part of the body, including the fuperior coverts of the tail, and the quills of the wings, brown; the fuperior coverts of the wings black, edged with bright yellow, which is confpicuous on the ground colour, and terminates each row of thefe by a waving line. The lower coverts of the tail are white; the tail, the bill, and the legs, are cinereous.

This bird is fmaller than the Lark; its total length is only fix inches and a half. I could not meafure it acrofs the wings; but thefe were certainly not broad, for when clofed they did not reach beyond the coverts of the tail. The bill and the legs are each eleven or twelve lines.

XXXIX. The

XXXIX.

The LITTLE RUFOUS-THROATED BROWN BLACKBIRD of Cayenne.

Turdus Pectoralis, Lath. Ind.
The Yellow-breasted Thrush, Lath.

I ſcarcely need add any thing to this deſcription. The rufous extends over the neck and breaſt; the bill is cinereous black, and the legs greeniſh-yellow. It is nearly of the ſize of the Goldfinch; its total length is hardly five inches, the bill ſeven or eight lines, the legs eight or nine; and the wings reach beyond the middle of the tail, which in all is only eighteen lines.

XL.

The OLIVE BLACKBIRD of St. Domingo.

Turdus Hiſpaniolenſis, Gmel.
Merula Olivacea Dominicenſis, Briſſ.
The Hiſpaniola Thruſh, Lath.

The upper-part of its body is olive, and the under gray, mixed confuſedly with the ſame colour. The inner webs of the tail-quills, of thoſe of the wings, and of the great coverts of these, are brown, edged with white or whitish; the bill and legs are grayish-brown.

This bird is scarcely larger than the Petty Chaps; its whole length is six inches, its alar extent eight and three-quarters, its tail two, its bill nine lines, its legs of the same length; its wings reach beyond the middle of its tail, which consists of twelve equal quills.

We may consider the *Olive Blackbird of Cayenne*, Pl. Enl. No. 558, as a variety of this; the only difference is, that the upper-part of the body is of a browner green, and the under of a lighter gray; the legs are also more blackish.

XLI.

The OLIVE BLACKBIRD of Barbary.

Mr. Bruce saw, in Barbary, a Blackbird, which was larger than the Missel; all the upper-part of the body was an olive yellow, the small coverts of the wings the same colour, with a tinge of brown, the great coverts and the quills black, the quills of the tail blackish, tipt with yellow, and all of equal length; the under-part of the body of a dirty white, the bill reddish brown, the legs short and lead-coloured; the wings reached only to the middle of the tail. It resembles much the Barbary Throstle already described,

described, but it has no speckles on its breast; and besides, there are other differences, which would lead us to refer them to two distinct species.

XLII.

The MOLOXITA, or the NUN of Abyssinia.

Turdus Monacha, Gmel.
The Nun Thrush, Lath.

Not only is this bird of the same figure and size with the Blackbirds, but like them it inhabits the forests, and lives on berries and fruits. Instinct, or perhaps experience, teaches it to lodge in trees near the brink of precipices: so that it is difficult to be shot, and still more to be found after it has dropped. It is remarkable for a great black cowl which covers the head and throat, and descends over the breast like a pointed stomacher; on this account it has been called the *Nun*. The whole of the upper-part of the body is yellow, more or less inclined to brown; the coverts of the wings, and the quills of the tail, are brown, edged with yellow; the quills of the wings different shades of black, and edged with light-gray or white; all the under-part of the body, and the thighs, light-brown; the legs cinereous, and the bill reddish.

XLIII.

The BLACK and WHITE BLACKBIRD of Abyſſinia.

Turdus Æthiopicus, Gmel.
The Ethiopian Thruſh, Lath.

Black extends over all the upper-part, from the bill incluſively to the end of the tail, excepting however the wings, on which we perceive a croſs bar of white, conſpicuous on the dark ground; white predominates in the under-part, and the legs are blackiſh. This bird is nearly of the ſize of the Red-Wing, but is rounder ſhaped; the tail is ſquare at the end, and the wings ſo ſhort, that they ſcarcely reach beyond its origin. It ſings nearly like the cuckoos, or rather the wooden clocks that imitate the cuckoos.

It haunts the thickeſt woods, and would be difficult to be diſcovered, but for its ſong; which would ſeem to ſhew that it does not ſeek ſafety in concealment, ſince the ſame inſtinct would enjoin ſilence.

This bird feeds on fruits and berries, like the Blackbirds and Thruſhes.

XLIV.

The BROWN BLACKBIRD of Abyssinia.

Turdus Abyssinicus, Gmel.
The Abyssinian Thrush, Lath.

The ancients have spoken of an Æthiopian olive-tree that bore no fruit: this bird feeds on the flower of that tree. If it were contented with that provision, few would have reason to complain. But it also eats grapes, and is very destructive in the season. This Blackbird is nearly as large as a Red-Wing; all the upper-part of the head and of the body is brown; the coverts of the wings of the same colour; the quills of the wings and of the tail deep-brown, edged with a lighter brown; the throat of a light brown; all the under-part of the body of a fulvous yellow, and the legs black*.

* Elegant drawings of the four last species were communicated by M. le Chevalier Bruce, (James Bruce, Esq. of Kinnaird,) of whom the author makes very honourable mention.

The

The GRISIN of Cayenne.

THE top of the head is blackiſh, the throat black; and this black colour extends from the eyes as far as the lower-part of the breaſt: it has a ſort of white eye-brows, which appear diſtinct on the duſky ground, and connect the eyes by a white line, which borders the baſe of the upper mandible. All the upper-part of the body is cinereous gray; the tail is deeper, and terminated with white; its lower coverts and the belly are alſo white; the coverts of the wings are blackiſh, and their limits accurately defined by a white border. The quills of the wings are edged without with light gray, and tipt with white; the bill is black, and the legs cinereous.

This bird is not larger than a Pettychaps; its length is about four inches and a half, its bill ſeven lines, its legs the ſame, and its wings reach to the middle of its tail, which is rather tapered.

In the female, the upper-part of the body is more cinereous than in the male; what is black in the latter is only blackiſh in the former, and for that reaſon the edge of the coverts of the wings is not ſo perceptible on the ground colour.

The VERDIN of Cochin China.

Turdus Cochinchinenſis, Gmel.
The Black-chinned Thruſh, Lath.

THE name of this bird ſufficiently marks its predominant colour. The green is ſhaded with a tinge of blue on the tail, on the outer edge of the great quills of the wings, and on the ſmall coverts near the back. The throat is velvet black, except the two ſmall blue ſpots which appear on both ſides of the lower mandible. This black extends behind the corners of the mouth, and riſes on the upper mandible, where it occupies the ſpace between its baſe and the eye, and below it is ſurrounded by a ſort of yellow high cape that falls on the breaſt; the belly is green, the bill black, and the legs blackiſh. This bird is nearly of the ſize of the Goldfinch. I could not meaſure its length, becauſe the tail was not fully grown when the bird was killed.

The bill is ten lines in length, and appears ſhaped like that of the Blackbirds, its edges being ſcalloped near the point. This little Blackbird is certainly a native of Cochin China, for it was found in the ſame box with the Muſk Animal, ſent directly from that country.

The AZURIN.

THIS bird is undoubtedly not a Blackbird, for it has neither the appearance nor the ſhape of one. However, there is ſome reſemblance in the form of the bill, the legs, &c. It has been called the *Guiana Blackbird.* I wait till travellers, ardent in the purſuit of Natural Hiſtory, make us acquainted with its true name, and, above all, with its habits. To judge from the little that is known of it, that is, from its external appearance, I ſhould range it between the Jays and the Blackbirds.

Three broad bars of fine velvet black, parted by two bars of orange-yellow, cover entirely the upper-part and the ſides of the head and of the neck. The throat is pure yellow, the breaſt decorated with a large blue plate; all the reſt of the lower-part of the body, including the inferior coverts of the tail, is radiated tranſverſely with theſe two laſt colours, and the blue appears alone on the quills of the tail, which are tapered. The upper-part of the body from the origin of the neck, and the neareſt coverts of the wings, are of a reddiſh brown; the moſt remote coverts are black, as are alſo the wing-quills: but ſome of the firſt have beſides a white ſpot, whence riſes a ſtripe of the ſame colour,

 deeply

deeply indented, and which runs almoſt parallel to the margin of the cloſed wing. The bill and legs are brown.

This bird is rather larger than a Blackbird; its whole length is eight inches and a half, its tail is two and a half, its bill twelve lines, and its legs eighteen. The wings, when cloſed, reach almoſt to the middle of the tail.

The SHORT TAIL.

Les Breves, Buff.

NATURE has eftablifhed important diftinctions between thefe birds and the Blackbirds; and I therefore do not hefitate to range them feparately. The fhortnefs of the tail, the thicknefs of the bill, and the length of the legs, are characteriftic features; and thefe muft involve other differences in their port, their habits, and perhaps in their difpofitions.

We are acquainted with only four birds of this fpecies; I fay fpecies, for the refemblance in the plumage is fo exact, that they muft be regarded as varieties only of a common ftem. In all of them the neck, the head, the tail, are black or partly black; the upper-part of the body is green of various intenfity; the fuperior coverts of the wings and tail are of a fine beryl colour, with a white or whitifh fpot on the great quills of the wing; laftly, in all, except that of the Philippines, the lower part of the body is yellow.

I. The SHORT-TAIL PHILIPPINE *. Its head and neck are covered with a fort of cowl entirely black, the tail of the fame colour; the under-part of the body, including the coverts and the fmall quills of the wings neareft the back, of a deep green; the breaft and the top of the

* *Corvus Brachyurus*, var. 1. Gmel. *Merula Viridis Atricapilla Moluccenfis*, Briff.

 the

the belly of a lighter green; the lower belly and the coverts of the tail of a roſe colour; the great quills of the wings black at their origin and at their extremity, and marked with a white ſpot between the two; the bill yellowiſh brown, and the legs orange.

The whole length of the bird is only ſix inches and a quarter, becauſe of its ſhort tail; but it is more than eight inches, when meaſured from the point of its bill to the end of its feet. It is nearly as large as the common Blackbird; its wings are twelve inches acroſs, and reach beyond the tail, which is only twelve lines long; the legs are eighteen.

II. The SHORT TAIL which Edwards has figured, Pl. 324, by the name of *Short-tailed Pie of the Eaſt-Indies* *. Its head is not entirely black; it has only three bars of that colour riſing from the baſe of the bill, the one ſtretching over the top of the head and behind the neck, and each of the others paſſing under the eye, and deſcending on the ſides of the neck. The two laſt bars are parted from the middle one by another bar, which is divided lengthwiſe by yellow and white; the yellow being contiguous to this middle bar, and the white contiguous to the black lateral bar. Alſo, the under-part of its tail and the lower belly are roſe-coloured, like the preceding; but all the reſt of the under-part of the body is yellow, the throat white, and the

* *Corvus Brachyurus*, var. 2. Gmel. *Coturnix Capenſis*, Klein. *The Madras Jay*, Ray. *The Bengal Quail*, Albin.

tail

tail edged with green at the end. It was brought from the iſland of Ceylon.

III. The SHORT TAIL of Bengal *. Like the firſt it has the head and neck covered with a black cowl, but on this two large orange-coloured eye-brows are diſtinctly defined: all the under-part of the body is yellow, and what was black in the great quills of the wing in the two preceding birds, is in this of a deep green, like the back. This bird is ſomewhat larger than the firſt, and of the ſize of an ordinary Blackbird.

IV. The SHORT TAIL of Madagaſcar †. The plumage of its head is alſo different from what we have juſt ſeen; the crown is of a blackiſh brown, which aſſumes a little yellow behind and on the ſides; the reſt is bounded by a half collar, which is black, and encircles the neck behind at its origin; and by two bars of the ſame colour, which riſing from the extremity of this half collar, paſs under the eyes, and terminate at the baſe of both mandibles; the tail is bordered at the end with a beryl colour; the wings are like thoſe of the firſt; the throat is mottled with white and yellow, and the under-part of the body is of a colour between yellow and brown.

* *Corvus Brachyurus*, Gmel. *Merula Viridis Moluccenſis*, Briſſ. *The Short-tailed Crow*, Lath.

† *Corvus Brachyurus*, var. 3. Gmel.

The MAINATE of the East Indies *

Gracula Religiosa, Linn. and Gmel.
Mainatus, Briss.
Minor Grakle, Lath.

THE slightest comparison will convince us, that this bird ought to be removed from the Blackbirds, Thrushes, Stares, and Jackdaws, with which it has been hastily ranged, and classed with the *Goulin* † of the Philippines, and especially with the *Martin* ‡, which belong to the same country, and have likewise naked spots on the head. This bird is scarcely larger than a common Blackbird; its plumage is entirely black, but more glossy on the upper-part of the body, the throat, the wings, and the tail, and has green and violet reflexions. What is most remarkable in the bird, is a double yellow comb, irregularly jagged, which rises on each side of the head, behind the eye; the two parts recline and approach each other, and on the back of the head they are parted only by a bar of long nar-

* It is the Indian Stare of Bontius, the *Corvus Javanensis* of Osbeck, and seems to be the *Merula Persica* of Camel, (Philos. Transact. No. 285.) This last author says, that "it is a sonorous and chattering bird, black, but naked about the eyes like "the *Illing*, but less so." A few lines below this, *Illing* is written *Iting*, which is our *Goulin*.

† *Bald Grakle,* Lath.

‡ *Paradise Grakle,* Lath.

row

N.° 81

THE MINOR GRAKLE

row feathers, which begins at the base of the bill; the other feathers on the crown of the head form a sort of black velvet. The tail, which is eighteen lines long, is yellow, but receives a reddish tinge near the tail; lastly, the legs are of an orange yellow. The tail of this bird is shorter, and the wings longer, than in the common Blackbird; these extend within half an inch of the end of the tail, and measure eighteen or twenty inches across. The tail consists of twelve quills, and of those of the wing, the first is the shortest, and the third the longest.

Such is the Mainate, No. 268, *Pl. Enl.*—But we must own that this species is subject to great variety, both in its plumage, in its size, and in the double comb which characterizes it. Before entering into detail, I shall mention that the Mainate has great talents for whistling, chanting, and even speaking; that its pronunciation is more free than that of the Parrot; that it has been called by distinction the Speaking Bird, and that its garrulity becomes troublesome *.

* Specific character of the *Gracula Religiosa*:—"It is violet "black, with a white spot on the wings, and a naked yellow stripe "on the back of the head." It is ten inches and a half long; lives upon fruits, cherries, grapes, &c.

VARI-

VARIETIES *of the* MAINATE.

I. The MAINATE of Briſſon. It differs from ours, becauſe it has on the middle of the firſt quills of the wing a white ſpot which does not appear in the coloured figure; whether that it did not exiſt in the ſubject, or eſcaped the deſigner: we may obſerve that the edge of the firſt quills is black, even where the white ſpot croſſes them.

II. The MAINATE of Bontius. Its plumage is blue of many tints, and conſequently ſomewhat different from that of ours, which is black, with reflexions of blue, green, violet, &c. Another remarkable difference is, that this blue ground was ſtrewed with ſpecks, like thoſe of the Stare, in point of ſhape and arrangement, but different with regard to colour; for Bontius ſubjoins that they are cinereous-gray.

III. The LITTLE MAINATE of Edwards. It has the white ſpot of Briſſon's on its wings; but what diſtinguiſhes it ſufficiently is, that the two creſts uniting behind the *occiput*, form a half crown, which ſtretches from one eye to the other. Edwards diſſected one, which was a female; and notwithſtanding the diſproportion in point of ſize, he leaves it to be decided, whether it was not a female of the following:

IV. The

IV. The GREAT MAINATE of Edwards *. Its creſt is the ſame as in the preceding, and it differs from that only in ſize, and in ſlight variations of colour. It is nearly the bulk of the Jay, and conſequently double the preceding, and the yellow of the bill and legs has no reddiſh tinge.—We are not informed whether the creſt of all theſe Mainates is ſubject to change of colour, according to the different ſeaſons of the year, and the various paſſions by which they are actuated.

* Gracula Religioſa, var. *Linn.* and *Gmel.* The Greater Minor, *Edw.* and *Lath.*

The GOULIN*

Gracula Calva, Linn. and Gmel.
Merula Calva Philippensis, Briss.
The Bald Grakle, Lath.

THERE are two ſpecimens of this ſpecies in the Royal Cabinet. In both, the upper-part of the body is of a light ſilver-gray, the tail and wings darker, each eye encircled by a bit of ſkin entirely bare, and forming an irregular ellipſe, inclined on its ſide, the eye being the inner focus: laſtly, on the crown of the head is a line of blackiſh feathers, which runs between theſe two ſkins; but one of theſe birds is much larger than the other. The largeſt is nearly of the bulk of the common Blackbird; the under-part of its body is brown, varied with ſome white ſpots, the naked ſkin which ſurrounds the eyes fleſh-coloured, the bill, the legs, and the nails, black. In the ſmaller, the under-part of the

* Camel, in the Philoſophical Tranſactions for 1703, ſays, that the *Goulin* is known in the Philippine iſlands by the names of *Iting*, *Illing*, and *Tabadura*. He adds, that it is a ſpecies of *Palalaca*, which is a Great Woodpecker. In this laſt aſſertion he is perhaps miſtaken; but there is little doubt that his *Gulin* or *Goulin* is the bird now under conſideration. His deſcription is as follows:—"It is of the bulk of the Stare; its bill, its wings, its tail, and "its legs, are black, the reſt ſilvery; the head is naked, except "a line of black feathers that runs on the crown. It ſings and "chatters much."

the body is of a yellowiſh brown; the bald parts of the head yellow, and alſo the legs, the nails, and the anterior part of the bill. Poivre informs us, that this naked ſkin, ſometimes yellow, ſometimes fleſh-coloured, which ſurrounds the eyes, is painted with a bright orange when the bird is angry; and this muſt probably happen likewiſe in the ſpring, when the bird burns with a paſſion as impetuous but more gentle. I retain the name of *Goulin*, which it receives in the Philippines, becauſe it is diſtinguiſhed from the Blackbird not only by the bald part on its head, but by the ſhape and thickneſs of its bill.

Sonnerat has brought from the Philippines a bald bird, which reſembles much the one figured No. 200, *Pl. Enl.* but differs in the ſize and plumage. It is near a foot in length; the two bits of naked ſkin which encircle the eyes are fleſh-coloured, and parted from the crown of the head by a line of black feathers, which runs between them. All the other feathers which ſurround this naked ſkin are alſo of fine black; and ſo is the under-part of the body, the wings, and the tail: the upper-part of the body is gray, but this colour is lighter on the rump and neck, deeper on the back and the loins. The bill is blackiſh; the wings very ſhort, and ſcarcely extend beyond the origin of the tail. If the two bald Blackbirds in the Royal Cabinet belong to the ſame ſpecies, we muſt regard the larger as a young ſubject, which had not attained its full growth,

growth, or received its true colours, and the ſmaller as one ſtill younger.

Theſe birds commonly neſtle in the holes of trees, eſpecially on the cocoa-nut tree; they live on fruits, and are very voracious, which has given riſe to the vulgar notion, that they have only one inteſtine, which extends ſtraight from the orifice of the ſtomach to the anus *.

* Specific character of the *Gracula Calva*:—" It is ſomewhat " aſh-coloured; its head naked on both ſides."

M

The PARADISE GRAKLE.

Le Martin, Buff.
Paradisea Tristis, Linn. and Gmel.
Gracula Tristis, Lath. Ind.
Merula Philippensis, Briss.

THIS bird feeds upon insects, and the havoc which it makes is the more considerable, as it has a gluttonous appetite: the various sorts of flies and caterpillars are its prey. Like the Carrion Crows and Magpies, it hovers about the horses, the oxen, and the hogs, in search of the vermin which often torment these animals to such a degree as to exhaust them, and even occasion death. The patient Quadrupedes are glad to get rid of these, and suffer, without molestation, often ten or twelve Paradise Grakles to perch on their back at once: but the intruders are not content with this indulgence; the skin need not be laid bare by some wound; the birds will peck with their bill into the raw flesh, and do more injury than the vermin which they extract. They may indeed be considered as carnivorous birds, whose prudence directs them to attack openly none but the weak and the feeble. A young one was known to seize a rat two inches long, exclusive of the tail, dash it repeatedly against the board of its cage, break the

 bones,

bones, and reduce every limb to a pliancy ſuited to its views; and then lay hold of it by the head, and almoſt in an inſtant ſwallow it entire. It reſted about a quarter of an hour to digeſt it, its wings drooping, and its air languid; but, after that interval, it ran with its uſual cheerfulneſs, and about an hour afterwards, having found another rat, it ſwallowed that as it did the firſt, and with as little inconvenience.

This bird is alſo very fond of graſshoppers; and as it deſtroys immenſe quantities, it is a valuable gueſt in countries curſed with theſe inſects, and it merits to have its hiſtory interwoven with that of man. It is found in India and the Philippines, and probably in the intermediate iſlands; but it has long been unknown in that of Bourbon. Not above twenty years ago, Deſforges-Boucher, Governor-general, and Poivre, the Intendant, perceiving this iſland deſolated by graſshoppers *, deliberated ſeriouſly about the means of extirpating theſe inſects; and for that purpoſe brought ſeveral pairs of Paradiſe Grakles from India, with the view to multiply them, and oppoſe them as auxiliaries to their formidable enemies. This plan promiſed to ſucceed; when unfortunately ſome of the coloniſts, noticing theſe birds eagerly boring in the new-ſown fields, fancied that they were ſearching for grain,

* Theſe graſshoppers had been introduced from Madagaſcar, their eggs being conveyed in the ſoil with ſome plants.

were inſtantly alarmed, and reported through the whole iſland that the Paradiſe Grakle was pernicious. The cauſe was conſidered in form: in defence of the birds it was urged, that they raked in new-ploughed grounds, not for the grain, but on account of the inſects, and were ſo far beneficial. However, they were proſcribed by the council, and two hours after the ſentence was paſſed, not one was to be found in the iſland. This prompt execution was followed by a ſpeedy repentance. The graſshoppers gained an aſcendency, and the people, who only view the preſent, regretted the loſs of the Paradiſe Grakles. De Morave, conſulting the inclinations of the ſettlers, procured four of theſe birds eight years after their proſcription. They were received with tranſports of joy. Their preſervation and breeding were made a ſtate affair; the laws held out to them protection, and the phyſicians on their part declared that their fleſh was unwholeſome. After ſo many and ſo powerful expedients, the deſired effect was produced; the Paradiſe Grakles multiplied, and the graſshoppers were entirely extirpated. But an oppoſite inconvenience has ariſen; the birds, ſupported no longer by inſects, have had recourſe to fruits, and have fed on the mulberries, grapes, and dates. They have even ſcraped up the grains of wheat, rice, maize, and beans; they have rifled the pigeon-houſes, and preyed on the young; and thus, after freeing the ſettlers from the

 graſs-

graſshoppers, they have themſelves become a more dreadful ſcourge *. Their rapid multiplication renders it difficult to ſtop their progreſs; unleſs perhaps a body of more powerful rapacious birds were employed againſt them; a plan which would ſoon be attended with other difficulties. The great ſecret would be to maintain a certain number of Paradiſe Grakles, and, at the ſame time, to contrive to reſtrain their farther multiplication. Perhaps an attentive obſervation of the nature and inſtincts of graſshoppers, would ſuggeſt a method of getting rid of them, without having recourſe to ſuch expenſive auxiliaries.

Theſe birds are not timorous, and are little diſturbed by the report of a muſket. They commonly take poſſeſſion of certain trees, or even certain rows of trees, often very near hamlets, to paſs the night. They alight in an evening in ſuch immenſe bodies, that the branches are entirely covered with them, and the leaves concealed. When thus aſſembled, they all begin to chatter together, and their noiſy ſociety is exceedingly troubleſome to their neighbours. Yet their natural ſong is pleaſant, varied, and extenſive. In the morning they diſperſe into the fields, either in ſmall flocks, or in pairs, according to the ſeaſon.

* They are ſtill more pernicious, as they devour the uſeful inſects, particularly that called the *Little Lion*, which preys upon the woolly lice that ſo much injure the coffee-ſhrubs.

They have two hatches in ſucceſſion every year, the firſt being in the middle of ſpring. Theſe turn out well, unleſs the ſeaſon be rainy. Their neſts are very rude, and they take no precaution to prevent the wet from penetrating. They faſten them in the leaves of the palm or other trees, and whenever an opportunity preſents, they prefer a hay-loft. Theſe birds are warmly attached to their young. When their neſts are about to be robbed, they flutter round, and utter a ſort of croaking, which indicates their rage, and dart upon the plunderer. Nor do their fruitleſs exertions extinguiſh their affection; they follow their brood, which, if ſet in a window or open place, the parents will carefully ſupply with food; nor will they in the leaſt be deterred by anxiety for their own ſafety.

The young Paradiſe Grakles are quickly trained, and eaſily learn to prattle. If kept in the poultry-yard, they ſpontaneouſly mimic the cries of all the domeſtic animals, hens, cocks, geeſe, dogs, ſheep, &c. and their chattering is accompanied with certain accents and geſtures, which are full of prettineſſes.

Theſe birds are rather larger than the Blackbirds; their bill and legs are yellow as in theſe, but longer, and the tail ſhorter. The head and neck are blackiſh; behind the eye is a naked reddiſh ſkin, of a triangular ſhape, the lower-part of the breaſt, and all the upper-part of the body, including the coverts of the wings and of the tail, of a chesnut brown; the belly white, the twelve quills of the tail, and the middle quills of the wings brown; the large ones blackish, from the tip to the middle, and thence to their origin white; which produces an oblong spot of that colour near the edge of each wing when it is closed; and in this situation the wings extend to two-thirds of the tail.

It is scarcely possible to distinguish the female from the male, by the external appearance *.

* The principal facts in the history of this bird were communicated by M. M. Sonnerat and De la Nux, correspondents of the Cabinet of Natural History.

M

No. 82

THE CHATTERER.

The CHATTERER*.

Le Jaseur, Buff.
Ampelis Garrulus, Linn. and Gmel.
Garrulus Bohemicus, Ray, Will. and Klein.
Bombycilla Bohemica, Briss.
Turdus Cristatus, Frisch.
The Silk Tail, Ray.
The Bohemian Chatterer, Penn. and Lath.

THIS bird is distinguished from all others by the small red appendices which terminate most of the middle quills of the wings; these appendices are nothing but the projection of

* In Aristotle, (lib. ix. 16.) Γναφαλος, which signifies a sort of matrass or pillow, alluding to the silky feathers of the Chatterer. Aldrovandus gives it the name *Ampelis*, from not the best authority, that of the poet Callimachus. This word, *Ampelis*, was besides applied to other small birds, such as the Beccafico, which, like the Chatterer, feeds upon grapes (Αμπελος denotes *a vine*). Some have reckoned it improperly the *Merops* of Aristotle, which is the *Bee-eater*: others have taken it for the *Avis Incendiaria* of the ancients, or the bird of the Hercynian forest, mentioned by Pliny; though its feathers do not cast fire at night, as alleged of that bird, unless this was a metaphorical allusion to the colour of the Chatterer's eyes, and the tears on its wings. This bird has also been called *Avis Bohemica*, *Adepellus*, *Pteroclia*, *Fullo*, *Gallulus-Sylvestris*, *Zinzirella*, which last is formed from its cry *zi*, *zi*, *ri*; and the German name *Zinzerelle* has the same derivation. In that language it is also termed *Boehmer*, *Boeheimle*, *Boehmische-Drostel*, *Hauben-Drostel* (hood thrush), *Pest-Vogel* (noxious bird), *Krieg-Vogel* (war-bird), *Wipstertz*, *Seide-Schwantz* (silk-tail), *Sehnee-Lesche* (snow quencher), *Schnee-Vogel*: in Swedish, *Siden-Swantz*: in Italian, *Beccofrisone*, *Galletto del Bosco*: in Bohemian, *Brkostaw*: in Polish, *Jedwabnicska*, *Jemiolucha*.

the ſhafts beyond the webs, which as they ſpread extend into the ſhape of a pallet, and aſſume a red colour. Sometimes as many have been reckoned as eight appendices on each ſide; ſome aſſert that the males have ſeven, and the females five; others that the females have none at all *. For my own part, I have ſeen ſpecimens which had ſeven in the one wing and five in the other; others which had only three; and others which had none; and at the ſame time exhibited other differences in the plumage. Laſtly, I have obſerved theſe appendices ſometimes parted longitudinally into two branches nearly equal, inſtead of forming as uſual the little pallets of a ſingle piece.

Linnæus has, with great propriety, ſeparated this bird from the Thruſhes and Blackbirds; obſerving, beſides the ſmall red appendices which diſtinguiſh it, that its proportions are different, its bill ſhorter, more hooked, and armed with a double tooth or ſcallop, which appears near the ends of both mandibles †. But it is not eaſy to conceive why he ſhould range it with the Shrikes, while he admits that it feeds on berries,

* Edwards.

† Dr. Liſter aſſerts that he obſerved, in one of theſe birds, that the edges of the upper mandible were not notched near the tip; this may be regarded as an individual variety: but the remark corrected a miſtake of the Doctor's, who had, like Linnæus ‡, claſſed the Chatterer with the Shrikes.

‡ In the *Fauna Suecica*, the Chatterer is termed *Lanius Garrulus*. T.

and

and is by no means carnivorous. There is indeed a conſiderable reſemblance between theſe and the Shrikes and Red-Backs, in the diſpoſition of the colours, particularly on the head, in the ſhape of the bill, &c.; but the difference of inſtinct is more important, and ought to preclude their aſſociation.

It is not eaſy to determine the native climate of this bird. We ſhould be much deceived, if from the names of Bohemian Jay, Bohemian Chatterer, &c. we inferred with Geſner, Briſſon, and others, that Bohemia is its principal abode. It only migrates thither, as into many other countries *. In Auſtria, it is conceived to be a native of Bohemia and of Stiria, becauſe it enters by the frontiers of theſe regions; but in Bohemia it might be called the bird of Saxony, and in Saxony the bird of Denmark, or of other countries on the ſhores of the Baltic. The Engliſh traders aſſured Dr. Liſter, that for near a century paſt the Chatterers were very common in Pruſſia. Rzaczynſki tells us, that they viſit Great and Little Poland and Lithuania. Reaumur was informed from Dreſden, that they breed in the neighbourhood of Peterſburg. Linnæus mentions, ſeemingly on good authority, that they ſpend the ſummer, and conſequently breed, in the countries beyond Sweden; but his correſpondents did not communicate the detail of

* Friſch.

circumſtances. Laſtly, Strahlemberg told Friſch that they are found in Tartary in the holes of the rocks, and no doubt they muſt build their neſts in theſe. But whatever be the country which the Chatterers chooſe for their reſidence, where they enjoy the ſweets of exiſtence, and tranſmit them to new generations, it is certain that they are not ſedentary, but make their excurſions over all Europe. They ſometimes appear in the North of England *, in France †, Italy ‡, and no doubt in Spain ; but this laſt is conjectural, for we muſt own that the natural hiſtory of this charming country, ſo rich and ſo contiguous, inhabited by a people ſo renowned, is not better known to us than that of California, or of Japan §.

The migrations of the Chatterers are in every country pretty regular with regard to the ſeaſon ; but if theſe be annual, as Aldrovandus was told, the birds by no means purſue con-

* The ſubject figured and deſcribed in the Britiſh Zoology was killed on Flamborough-Moor, Yorkſhire. The two which Dr. Liſter ſaw were killed near the city of York. *See* Philoſophical Tranſactions, No. 173, Art. 3.

† A few years ſince a Chatterer was killed at Marcilly near the Ferté-Lowendhal; and lately four were taken at Beauce in the heart of winter, which had taken ſhelter in a pigeon-houſe. SALERNE.

‡ Aldrovandus.

§ Bowles' Natural Hiſtory of Spain has been ſince publiſhed. T.

Geſner ſays, that he never ſaw the Chatterer, aud that it is almoſt every where very rare. We may at leaſt conclude that it is rare in Switzerland.

ſtantly

ſtantly the ſame route. The young Prince Adam d'Averſperg, Chamberlain of their Imperial Majeſties, and one of the Lords of Bohemia, who poſſeſſes the moſt extenſive chaſe-grounds, and makes the nobleſt uſe of them, ſince he aims at the progreſs of Natural Hiſtory, informs us, in a Memoir addreſſed to the Count de Buffon *, that this bird flits every three or four years † from the mountains of Bohemia and Stiria, into Auſtria, in the beginning of the autumn; that it returns about the end of that ſeaſon; and that, even in Bohemia, not one is ſeen during the winter. However, it is ſaid, in Sileſia, the winter is the time when theſe birds are found on mountains. Thoſe which have ſtrayed into France and England appeared in the depth of the winter, but always in ſmall numbers: a circumſtance which would ſhew that they were parted from the great body by ſome accident, and too much fatigued, or too young to recover their route. We might alſo infer, that France, England, and even Sweden, are not ſituated in the courſe of the principal migration; but we cannot draw the ſame concluſion with regard to Italy, for immenſe numbers of theſe birds have there been ſeveral times obſerved to arrive. This was particularly the

* This Prince ſent with his Memoir a ſtuffed Chatterer from his collection, and preſented it to the Royal Cabinet.

† Others ſay every five years, and others every ſeven years. Gesner.

caſe

case in 1571, in the month of December; at that time it was not uncommon to see flocks of an hundred or more, and forty were often caught at once. The same event took place in February 1530 *, when Charles V. caused himself to be crowned at Bologna; for in countries where these birds appear at distant intervals, their visits form an epoch in political history, especially since when they are very numerous, they announce to the frightened imaginations of the people war and pestilence. From these calamities we must, however, except that of earthquakes; for in 1551, when the Chatterers again appeared, it was observed that they spread through Modena, Placentia, and almost all parts of Italy †, but constantly avoided Ferrara, as if they had a *presentiment* of the earthquake which happened soon after, and dispersed even the birds of that country.

We cannot well assign the cause that determines these birds to leave their ordinary residence, and to roam into distant climes. It is not excessive cold; for they are embodied for their expedition as early as the beginning of autumn; and besides, their migration is only occasional, happening once in three or four years, or only after six or seven years, and their nume-

* As Italy is warmer than Germany, they might appear there later, and I doubt not but in more northern countries they remain a great part of the winter in years when it is not rigorous.

† Aldrovandus.

rous hoſt often darkens the heavens *. Are we to aſcribe theſe migrations to their prodigious multiplication; like the locuſts, and thoſe rats of the north called *lemings*; and, as has happened to the human ſpecies, when they were leſs civilized, and conſequently ſtronger, and more independent of the equilibrium which at length eſtabliſhes itſelf between all the powers of nature? Or are the Chatterers compelled repeatedly by local ſcarcity to quit their abodes, and ſeek ſubſiſtence in other climates? It is ſaid that they penetrate far into the arctic regions; and this is confirmed by Count Strahlenberg, who, as we have already noticed, ſaw them in Tartary.

When the Chatterer reſides in wine countries, it ſeems fondeſt of grapes; whence Aldrovandus calls it *Ampelis*, or *Vine Bird*. Next to theſe, it feeds upon the berries of privet, of bramble, of juniper, of laurel; upon almonds, apples, ſorbs, wild gooſeberries, figs, and, in general, upon melting juicy fruits. The one which Aldrovandus kept near three months, would not eat ivy-berries or raw fleſh till driven to extremity, and never touched grain. That which they tried to breed in the *menagerie* at Vienna was fed upon crumbs of white bread, maſhed carrots, bruiſed hemp-ſeed, and juniper berries,

* Geſner.

which

which it preferred *; but in ſpite of all the care which was taken, it lived only five or ſix days. Not that the Chatterer is difficult to tame, but that a bird, which has roamed at will, and been accuſtomed to provide its own ſubſiſtence, thrives always beſt in the ſtate of liberty. Reaumur remarks, that the Chatterers love cleanlineſs, and when confined they conſtantly void their excrements in the ſame ſpot †.

Theſe birds are entirely of a ſocial diſpoſition; they commonly fly in flocks, and ſometimes form immenſe bodies. Beſide this general amicable turn, and independent of their attachment to the females, they are ſuſceptible of particular friendſhips to individuals of the ſame ſex. But their affectionate temper, which implies more gentleneſs than activity, more ſecurity than diſcernment, more ſimplicity than prudence, more ſenſibility than vigour, precipitates them oftener into danger than ſuch as are more ſelfiſh. Accordingly, theſe birds are reckoned the moſt ſtupid, and they are caught in the greateſt numbers. They are generally taken with the Thruſh, which migrates about the ſame time, and their fleſh has nearly the ſame taſte ‡; which is natural to ſuppoſe, ſince they live upon the ſame food.

* Memoire of the Prince d'Averſperg.

† *See* Salerne, p. 253,

‡ Geſner tells us that their fleſh is very delicate, ſerved up at the beſt tables, and the liver eſpecially highly valued. The Prince d'Averſperg aſſures us, that it is better taſted than that of the Thruſh and

food. I ſhall add, that many of them are killed at once, for they ſit cloſe together *.

They utter their cry as they riſe from the ground; this is *zi*, *zi*, *ri*, according to Friſch, and all thoſe who have ſeen them alive; it is rather a chirrup than a ſong, and hence their name of *Chatterer*. Reaumur will not even admit that they can chant; but Prince d'Averſperg ſays that their notes are very pleaſant. Perhaps, in countries where they breed, they may warble in the ſeaſon of love, while they only chirp or chatter in other places; and when confined in cages they may be totally ſilent.

The plumage is agreeable when the bird is ſtill; but when it diſplays its wings, expands its tail, and erects its creſt, in the act of flying, its appearance is charming. Its eyes, which are of a beautiful red, ſhine with uncommon luſtre in the middle of the black band, in which they are placed. This black extends under the throat, and quite round the bill; the different ſhades of wine colour on its head, back, and breaſt, and the aſh colour of the rump, are ſurrounded with a frame enamelled with white, with yellow, and with red, formed by the different ſpots of the wings and tail: the latter is cinereous at its origin, blackiſh in its middle, and yellow at its end:

and Blackbird. On the other hand, Schwenckfeld ſays that it is very indifferent eating, and unwholeſome. This muſt depend much on the quality of the ſubſtances on which the bird feeds.

* Friſch.

end: the quills of the wings are blackiſh, the third and fourth are marked with white near the tip, the five following marked with yellow, and moſt of theſe terminated with broad tears of a red colour, of which I have ſpoken in the beginning of this article. The bill and legs are black, and ſhorter in proportion than in the Blackbird. The total length of the bird is, according to Briſſon, ſeven inches and three-quarters, its tail two and a quarter, its bill nine lines, its legs the ſame, and its alar extent thirteen inches. For my part, I have obſerved that the dimenſions were all greater than here ſtated; owing, perhaps, to difference of age or ſex, or even between the individuals.

I am not acquainted with the plumage of the young Chatterers, but Aldrovandus tells us that the margin of the tail is of a duller yellow in the females, and that the middle quills have whitiſh marks, and not yellow, as in the males. He adds a circumſtance which is hard to believe, though he aſſerts it from his own obſervation, that in the females the tail conſiſts of twelve quills, but in the males of ten only. It is much more credible that the male ſpecimens examined by Aldrovandus had loſt two of their quills *.

* Specific character of the *Ampelis Garrulus*:—"The back of "its head is creſted, the ſecondary wing-quills are membranous, "and coloured at the tip."

VARI-

VARIETY *of the* CHATTERER.

We may obſerve that the Chatterer is proportionably much broader acroſs the wings than the Blackbird or Thruſhes. Aldrovandus has alſo remarked, that the *ſternum* is of a ſhape better adapted for cutting the air, and accelerating its courſe. We need not then be ſurpriſed that it performs ſuch diſtant journies in Europe; and ſince it ſpends the ſummer in the countries of the north, we ſhould naturally expect to diſcover it in America. And this is actually the caſe. Reaumur received ſeveral from Canada, where they were called *Recollet* *, on account of the reſemblance perceived between the creſt and a monk's frock. From Canada they could eaſily ſpread into the ſouthern colonies. Cateſby deſcribes them among the birds of Carolina: Fernandez ſaw them in Mexico near Tezcuco †: I have examined ſome which were ſent from Cayenne. This bird is not above an ounce in weight, according to Cateſby; its creſt, when erected, is pyramidal, its bill is black, with a large opening, its eyes placed on a bar of the ſame colour, ſeparated from the ground by two white ſtreaks, the extremity of the tail edged

* It is the *Caquantototl* of Fernandez.

† He ſays that it delights to dwell in the mountains, that it lives on ſmall ſeeds, that its ſong is ordinary, that its fleſh is indifferent food.

with

with a ſhining white, the upper-part of the head, the throat, and the back, hazel, with a wine-tinge; the coverts and quills of the wings, the lower-part of the back, the rump, and a great part of the tail, of different ſhades of cinereous; the breaſt, and the inferior coverts of the tail, whitiſh; the belly and flanks of a pale yellow. It appears from this deſcription, and from the meaſures which have been taken, that the American Chatterer is rather ſmaller than the European ſort; that its wings have leſs of the enamel, and are rather of a duſkier hue; and that the wings do not extend ſo far in proportion as the tail. But it is undoubtedly the ſame ſpecies; for ſeven or eight middle quills of its wing are terminated by the little red appendices. Brooke, ſurgeon in Maryland, told Edwards, that the females wanted theſe appendices, and that the colours of their plumage were not ſo bright as thoſe of the males. The Cayenne Chatterers which I examined had really not theſe appendices, and the ſhades of the plumage were in general fainter, as it commonly happens in the females.

N.° 83

FIG. 1. THE HAWFINCH. FIG. 2. THE COMMON CROSSBILL.

The GROSBEAK*.

Le Gros-Bec, Buff.
Loxia-Coccothrauſtes, Linn. and Gmel.
Coccothrauſtes, Geſner, Aldrov. Briſſ. &c.
The Groſbeak, or *Hawſinch*, Will. Edw. &c.

THIS bird is an inhabitant of the temperate climates, from Spain and Italy, as far as Sweden. The ſpecies, though rather ſtationary, is not numerous. It appears every year in ſome of the provinces of France, and leaves them only for a ſhort time in the ſevereſt winters †. It generally inhabits in the woods during the ſummer, and ſometimes the vineyards; and in winter it reſorts near the hamlets and farms. It is a ſilent bird which is ſeldom heard, and ſeems

* Its Greek and Latin name *Coccothrauſtes* is derived from κοκκος, a *grain* or *kernel*, and θραυω, to *break*, becauſe it feeds upon cherry-ſtones: However, that appellation might have been given to ſome other bird that had the ſame habit; for Heſychius and Varro, the only ancient authors in whoſe writings the word is found, ſay no more than that "the *Coccothrauſtes* is a certain bird." In Italy it is called *Froſone, Friſone, Griſone, Franguet del Re, Franguet Montano*: in Germany, *Heine-Byſſer, Bollebiek, Kirſch-Finck, Kern-Beiſz, Riſch Leſke*: in Switzerland, *Klepper*: in Sweden, *Talbin*.

† It is difficult to reconcile this remark, of which I am certain, with the account given by the authors of the Britiſh Zoology, that the Groſbeak is ſeldom ſeen in England, and never except in winter. Perhaps, as there are few foreſts in England, there are alſo few of theſe birds, which reſide only in the woods; and as they approach the hamlets only in winter, obſervers may never have ſeen them but in that ſeaſon.

 to

to have no fong or decided warble. Nor is its organ of hearing fo perfect as that of other birds, for though it refides in the forefts, it cannot be enticed by the call. Gefner, and moft naturalifts after him, have faid, that the Grofbeak is good eating. I have tafted the flefh, but it feemed neither pleafant nor juicy.

I have obferved in Burgundy that thefe birds are much fewer in winter than in fummer, and that great numbers of them arrive about the 10th of April in fmall flocks, and perch among the copfes, building their nefts * on trees, generally at the height of ten or twelve feet, where the boughs divide from the trunk. The materials are, like thofe of the Turtle, dry fticks, matted with fmall roots. They commonly lay five bluifh eggs fpotted with brown. We might fuppofe that they breed only once a-year, fince the fpecies is not numerous. They feed their young with infects, chryfalids, &c.; and when they are about to be robbed of their family, they make a vigorous defence, and bite fiercely. Their thick ftrong bill enables them to crack nuts, and other hard fubftances; and though

* A Grofbeak's neft was found the 24th of April 1774, on a plum-tree ten or twelve feet high, in the fork of a branch; it was of a round hemifpherical fhape, compofed externally with fmall roots and fome lichens, and internally with other fmall roots more flender; it contained four eggs fomewhat pointed, their great diameter nine or ten lines, their fmall diameter fix lines; they were marked with fpots of an olive brown, and with irregular blackifh ftreaks faintly impreffed on a ground of bluifh light-green. Note communicated by *M. Gueneau de Montbeillard.*

they are granivorous, they alſo live much upon inſects. I have kept them a long time in voleries; they reject fleſh, but readily eat any thing elſe. They muſt be confined in a ſeparate cage, for without ſeeming at all diſcompoſed, or making the leaſt noiſe, they kill the weaker birds that are lodged with them. They attack, not by ſtriking with the point of the bill, but by biting out a morſel of the ſkin. When at liberty, they live upon all ſorts of grain, and kernels of fruits; the Orioles eat the pulp of cherries, but the Groſbeaks break them to obtain the kernel; they feed alſo on fir and pine cones, and on beech maſt, &c.

This bird is ſolitary, ſhy, and ſilent; its ear is inſenſible, and its prolific powers are inferior to thoſe of moſt other birds. It ſeems to have its qualities concentrated in itſelf, and is not ſubject to any of the varieties which almoſt all proceed from the luxuriance of nature. The male and female are of the ſame ſize, and much reſemble each other. The ſpecies is uniform in our climate; but in foreign countries there exiſt many analogous birds, which ſhall be enumerated in the ſucceeding article*. [A]

* The upper-mandible is cinereous, but of a lighter tint near the baſe; the lower-mandible is cinereous at the edges which cloſe into the upper; its under ſide is fleſh-coloured, with a cinereous caſt. The tongue is fleſhy, ſmall, and pointed; the gizzard is very muſcular, preceded by a pouch, containing in ſummer bruiſed hemp ſeeds, green caterpillars almoſt entire, and very ſmall

ſmall ſtones. In a ſubject which I diſſected lately, the inteſtinal tube from the *pharynx* to the craw was three inches and an half long, and from the gizzard to the *anus* about a foot. It had no *cæcum* or gall-bladder. *Obſervations communicated by M. Gueneau de Montbeillard, the* 22*d April* 1774.

[A] Specific character of the *Loxia Coccothrauſtes* :—" It has a " white line on the wings, the middle quills of the wings are " rhomboid-ſhaped at the tips, the quills of the tail are black on " the thinner ſide of the baſe."

The CROSSBILL*.

Le Bec Croisé, Buff.
Loxia Curvirostra, Linn. and Gmel.
Loxia, Gesner, Aldrov. Briss. &c.
The Shell-Apple, or *Crossbill*, Will. Edw. &c.

THE species of the Crossbill is closely related to that of the Grosbeak. Both have the same size, the same figure, the same instincts †. The Crossbill is distinguished only by a sort of deformity in its bill, a character, or rather a defect, which belongs to it alone of all the winged tribe. What proves that it is a defect, an error of nature rather than a permanent feature, is, that it is variable; the bill in some subjects crosses to the left, in others to the right; but the productions of nature are regular in their developement, and uniform in their arrangement. I should therefore impute this difference of position to the way in which the bird has used its bill, according as it has been more accustomed to employ the one side or the other to lay hold of its food. The same takes place in men, who,

* Gesner gave it the name *Loxia*, from the Greek λοξος, *oblique*, on account of the crossing of its bill. In Germany it is called *Kreutz-Schnabel* (Crossbill), *Kreutz-Vogel*: in Poland, *Rzywonos*: in Sweden, *Korsnaef*, *Kiaegelrifware*.

† Frisch conceives them to be so nearly allied, that they would breed together.

from

from habit, prefer the right hand to the left *. Each mandible of the Crofsbill is affected by an exuberance of growth, fo that in time the two points are parted afunder, and the bird can take its food only by the fide; and hence if it oftener ufes the left, the bill will protrude to the right, and *vice verfâ*.

But every thing has its utility, and each fentient being learns to draw advantage even from its defects. This bill, hooked upwards and downwards, and bent in oppofite directions, feems to have been formed for the purpofe of detaching the fcales of fir-cones, and obtaining the feeds lodged beneath thefe, which are the principal food of the bird. It raifes each fcale with its lower mandible, and breaks it off with the upper; it may be obferved to perform this manœuvre in its cage. This bill alfo affifts its owner in climbing, and it dextroufly mounts from the lower to the upper bars of its cage. From its mode of fcrambling, and the beauty of its colours, it has been called by fome *the German Parrot*.

* This obfervation muft be qualified. If habit were the fole caufe of this difference, as many people would be left-handed as right-handed. But the number of the former is very fmall, compared with that of the latter, and therefore the right-hand muft by original conftitution be ftronger than the left, however much the difference is afterwards increafed from habit. The fame reafoning feems applicable to the Crofsbill. T.

The

The Crofsbill inhabits only the cold climates, or the mountains in temperate countries. It is found in Sweden, in Poland, in Germany, in Switzerland, and among the Alps and Pyrenees. It is quite ftationary in countries where it lives the whole year; but fometimes it accidentally appears in large flocks in other regions. In 1756 and 1757, great numbers were feen in the neighbourhood of London. They do not arrive at ftated feafons, but feem to be rather directed by chance, and many years pafs without their being at all obferved. The Nut-Crackers, and fome other birds, are fubject to the fame irregular migrations, which occur only once in twenty or thirty years. The only caufe which can be affigned is, that they have been deprived of their ufual fubfiftence in the climates where they inhabit, by the inclemency of the feafon; or have been driven upon the coaft by the violence of a ftorm or hurricane: for they arrive in fuch numbers, and appear fo much exhaufted, that they are carelefs of their exiftence, and allow themfelves to be caught by the hand.

We might prefume that the fpecies of the Crofsbill, which prefers the cold climates, would be found in the north of the New Continent, as in that of the Old: yet no traveller to America has taken notice of it. But befides the general prefumption which is verified by analogy, there is a fact which feems to prove our opinion; the

 Crofsbill

Crofsbill is found in Greenland, whence it was brought to Edwards by the whale-fifhers; and that naturalift, who was better acquainted than any perfon with the nature of birds, remarks properly, that both the land and the water fort which inhabit the arctic regions, appear indifferently in the north of America or of Europe.

The Crofsbill is one of thofe birds whofe colours are the moft fubject to vary; among a great number we can fcarcely find two individuals that are exactly fimilar; not only are the fhades of the plumage different, but the pofition of the colours change with the feafon and the age. Edwards, who examined a prodigious number of them, and fought to mark the limits of variation, paints the male with a rofe colour, and the female with a yellowifh green; but in both, the bill, the eyes, the thighs, and the legs, are precifely the fame in regard to fhape and colours. Gefner tells us that he kept one of thefe birds, which was blackifh in September, and affumed a red colour in October. He adds, that the parts where the red began to appear, were the under-fide of the neck, the breaft, and the belly; that this red afterwards became yellow, and that winter efpecially is the feafon when thefe changes take place, and that, at different times, it is faid they receive a red, yellow, green, and cinereous caft. We muft not, therefore, with our modern nomenclators, reckon as a feparate fpecies, or a particular variety, a

greenifh

greenish Crossbill *, found in the Pyrenees, since it occurs equally in other places; and in certain seasons it has in all countries that colour. According to Frisch, who was perfectly acquainted with these birds, which are common in Germany, the colour of the adult male is reddish, or green mixed with red; but they lose this red, like the Linnets, when they are kept in the cage, and only retain the green, which is more deeply impressed both in the young and in the old. For this reason they are called in some parts of Germany *krinis* or *grünitz*, that is, greenish bird. The two extreme colours have not therefore been well chosen by Edwards; we must not infer, as his figures would suggest, that the male is red, and the female green; there is every reason to believe, that in the same season, and at the same age, the female differs from the male only in the greater faintness of the colours.

This bird, which is so analogous to the Grosbeak, resembles it also in stupidity. One may approach it, fire upon it without scaring it, and sometimes even catch it by the hand; and as it is equally inactive and secure, it falls an easy victim to all the birds of prey. It is mute in summer, and its feeble notes are only heard in winter †. It is quite placid in captivity, and

* *Loxia Pyrenaica*, Barrere. *Loxia Rufescens*, Brisson.

† Gesner.

lives

lives long in a cage. It is fed with bruiſed hemp-ſeed, and this contributes to make it ſooner loſe its red *. In ſummer, its fleſh is ſaid to be good eating †.

Theſe birds delight only in the dark foreſts of pines and firs, and ſeem to dread the effulgence of day. Nor do they yield to the genial influence of the ſeaſons; it is not in ſpring, but in the depth of winter, that their loves commence. They build as early as January, and their young are grown before the other birds begin to lay. They place their neſts under the large branches of the pine, fixing them with the reſin of that tree, and beſmearing them with that ſubſtance, ſo that the melted ſnow or the rains cannot penetrate. In the young, as in thoſe of other birds, the bill, or rather corners of its opening, are yellow, and they hold it always open as long as they are fed by the mother. We are not told how many eggs they lay, but we may preſume, from their ſize and their reſemblance to the Groſbeak, that the number is four or five, and that they hatch only once a-year. [A]

* Friſch. † Geſner and Friſch.

[A] Specific character of the Croſsbill, *Loxia Curviroſtra*, LINN.—" It is red, its bill forked." It is of the ſize of a Lark, being ſix inches and a half long.

FOREIGN BIRDS,

THAT ARE RELATED TO THE GROSBEAK.

I.

THE East-India bird, delineated in the *Pl. Enl.* No. 101, fig. 1. under the name of *Coromandel Grosbeak*, and which name we have still retained, because it appears to be the same species with that of Europe. The shape, the size, the bill, the length of the tail, are the same in both, and the only difference consists in the colours, which are also disposed in the same order. In short, we may impute the difference of shade to the influence of climate, and consider this Coromandel bird, which no naturalist has taken notice of, as a beautiful variety of the European Grosbeak.

II.

The American bird, No. 154, *Pl. Enl.* termed *the Blue American Grosbeak*, on which we have bestowed no discriminating name, because we are

are not certain if it is a peculiar ſpecies, different from that of Europe; for in ſize and figure it is the ſame with our Groſbeak. The only difference is, that it has more red on its bill, and more blue in its plumage; and if its tail were not longer, we ſhould not heſitate to pronounce that it is a mere variety, occaſioned by the influence of climate. No naturaliſt has noticed this new variety or ſpecies, which we muſt be careful not to confound with the Carolina bird, called by Cateſby *the Blue Groſbeak.*

III.

The HARD-BILL.

Le Dur-Bec, Buff.
Loxia Enucleator, Linn. and Gmel.
Coccothrauſtes Canadenſis, Briſſ.
The Greateſt Bulfinch, Edw.
The Pine Groſbeak, Penn. and Lath.

The Canada bird, delineated *Pl. Enl.* No. 135, fig. 1. under the name of *Canada Groſbeak*, and which we have called *Hard-bill*, becauſe its bill is comparatively harder, ſhorter, and ſtronger, than in the others; and it was proper to apply to it a diſtinct name, ſince it differs not only from the European Groſbeaks, but from all thoſe of America and of other climates. It is of a beautiful red, as large as our Groſbeak,

 but

but longer tailed, and may be eaſily diſtinguiſhed from all the other birds by the inſpection of the coloured figure. The female, has only a little reddiſh on its head and rump,, and a ſlight tinge of roſe-colour on the lower-part of its body. Salerne tells us, that in Canada this bird is called *bouvreuil (Bulfinch)*. This name has not been ill applied, for there is perhaps an affinity between it and the Bulfinch. The inhabitants of that part of America could decide this point by a very ſimple obſervation, *viz.* by noticing whether it whiſtles almoſt continually like the Bulfinch, or is almoſt mute like the Groſbeak *.

* Specific character of the *Loxia Enucleator*:—"It has a double white line on the wings, and all the quills of its tail are "blackiſh."—It is about nine inches long. Found in all the northern parts of America, from Canada to the weſtern ſide of the continent. It arrives in Hudſon's Bay early in the ſpring; lodges among the pines and junipers; builds its neſt at a ſmall height from the ground; lays four eggs, which it hatches in June. Theſe birds occur alſo in the north of Europe and Aſia. They are frequent in Ruſſia and Siberia; and Mr. Pennant tells us that he ſaw them in the pine foreſts near Invercauld, Aberdeenſhire, in the month of Auguſt.

IV. The

IV.

The CRESTED CARDINAL.

Le Cardiual Huppé, } Linn. and Gmel.
Loxia Cardinalis, }
Coccothrauſtes Virginiana, Briſſ.
Coccothrauſtes Indica Criſtata, Ray, and Will.
The Red Groſbeak, Albin.
The Red Bird, Kalm's Travels.
The Virginia Nightingale, Will.
The Cardinal Croſbeak, Penn. and Lath.

This is a native of the temperate climates of America, and figured No. 37, *Pl. Enl.* by the name of the *Virginia Groſbeak.* It is alſo called the *Creſted Cardinal*, which name we retain, as denoting its two characters, its colour, and its creſt. This bird reſembles much the Pine Groſbeak; the ſize, and, in a great meaſure, the plumage, are the ſame; the bill is as ſtrong, the tail of the ſame length, and the climate is nearly the ſame. We might, therefore, but for the creſt, reckon it a variety of that beautiful ſpecies. The colours in the male are much brighter than in the female, whoſe plumage is not red, but only reddiſh-brown; its bill is alſo of a much fainter red, though both have the creſt. I ſhould range this bird rather with the Bulfinch and the Chaffinch, than with the Groſbeak, ſince it ſings agreeably; whereas the Groſbeak is ſilent. Salerne ſays, that the warble of the

No. 84

THE CRESTED CARDINAL.

the Crefted Cardinal is charming, and refembles the fong of the Nightingale; and that it can be taught alfo to fpeak like the Canary birds. He adds, that this bird, which he obferved alive, is bold, ftrong, and vigorous, that it feeds upon feeds, particularly thofe of millet, and is eafily tamed *.

* Specific character of the *Loxia Cardinalis*:—" Its head is " red, its bridle black, its bill and legs blood-coloured." It is nine inches long: is found through the whole extent of North America. It feeds chiefly on Indian corn, of which it makes a provifion for the winter, artfully concealing the depofit with leaves and fmall branches. It warbles delightfully in the fpring mornings on the fummits of the loftieft trees. Its fong is faid to refemble that of the Throftle. It is a hardy and familiar bird.

The four birds which we have juft mentioned are all nearly of the fame fize with the European Grofbeak. But there are many other intermediate or fmaller fpecies, which we fhall range according to their fize and climate, and which, though all different from each other, may beft be compared with the Grofbeaks, to which they are more analogous than to any other. We may name them the *Middle Grofbeaks* and the *Little Grofbeaks*.

V. The

V.

The ROSE-THROAT.

Loxia Ludoviciana, Linn. and Gmel.
Coccothrauſtes Ludoviciana, Briſſ.
The Red-breaſted Groſbeak, Penn. and Lath.

The firſt of theſe ſpecies of the middle ſize is that of the *Pl. Enl.* No. 153, fig. 2. termed *the Groſbeak of Louiſiana.* Its throat is of a fine red roſe colour, and differs ſo much from all other ſpecies of the ſame genus, that it merits a diſtinct name. Briſſon firſt mentioned this bird, and has given a tolerably good figure of it; but he ſays nothing of its habits. The ſettlers in Louiſiana could inform us *.

* Specific character of the *Loxia Ludoviciana* :—" It is black; " its breaſt, its belly, the ſtripe on its wings, and the baſe of its " tail-quills, are white." The female is ſpotted with white on the head. It inhabits the greater part of North America: In the State of New-York it is reckoned a ſcarce bird. It appears there in May, and retires in Auguſt.

VI.

The GRIVELIN.

Loxia Braſiliana, Lath.

The ſecond ſpecies of the middling Groſbeaks is Fig. 1. No. 309, *Pl. Enl.* and there termed

termed the *Brazilian Grosbeak*. We have given it the name of *Grivelin*, because the under-part of its body is speckled like as in the Thrushes *(grives)*. As it is a beautiful bird, and unlike any other, it merited an appropriated name. It seems to be much related to the bird mentioned by Marcgrave, and which is called in Brazil *Guira*, *Tirica*. However, as the short description given by that author does not exactly correspond with our Grivelin, we cannot decide with regard to the identity of the species.

These middle-sized species, and those still smaller, are much more like the Sparrow in point of bulk and shape; but we have allowed them to remain with the Grosbeak, because their bill resembles that of these birds, and is much broader at the base than that of the Sparrow.

VII.

The RED BLACK.

The third species of the middle-sized Grosbeak is the bird delineated Fig. 2. No. 309. *Pl. Enl.* under the name of *the Cayenne Grosbeak*. We have called it the *Red Black*, because the whole of its body is red, and the breast and belly black. This bird, which is brought from Cayenne, has been noticed by no naturalist; but as we did not ſee it alive, we cannot deſcribe its habits. The people of Guiana could inſtruct us in that point.

VIII.

The FLAVERT*.

Loxia Canadenſis, Linn. and Gmel.
Coccothrauſtes Cayanenſis, Briſſ.
The Canada Groſbeak, Penn. and Lath.

The fourth ſpecies of theſe foreign middle-ſized Groſbeaks is the bird Fig. 2. No. 152. *Pl. Enl.* termed *the Cayenne Groſbeak*. It is yellow and green, and therefore differs from the preceding almoſt as much as it can do with regard to colours; but as its ſize, the ſhape of its body and of its bill, and its climate, are the ſame, we may reckon it a ſpecies cloſely related to the Red Black, if it be not a variety ariſing merely from the difference of age or ſex. Briſſon is the firſt who took notice of it. [A]

* *i. e.* The yellow-green.

[A] Specific character of the *Loxia Canadenſis*:—"It is olive-"green, above olive-yellow, its bridle black."

IX. The

IX.

The FAN-TAILED GROSBEAK.

La Queue en Eventail, Buff.
Loxia Flabellifera, Gmel.

The fifth ſpecies of theſe birds is that figured *Pl. Enl.* No. 380. under the name of *the Fan-tail of Virginia*. We received it from that part of America, and it has not been noticed by any preceding author. The upper figure, No. 380. repreſents probably the male, and the under the female, for its colours are not ſo vivid. We received theſe birds alive, but not being able to preſerve them, we could not decide whether we ſhould attribute the differences to ſex or to age. They are ſo remarkable for the ſhape of their tail, which is expanded horizontally, that this character alone is ſufficient to diſtinguiſh them from others of the ſame genus *.

* Specific character of the *Loxia Flabellifera*:—"It is duſky-"red, which below is more dilute; its bill, its wing-quills, its "tail, and its legs, are black." It is five inches long.

X. The

X.

The PADDA, or RICE-BIRD.

Loxia Oryzivora, Linn. and Gmel.
Coccothrauſtes Sinenſis Cinerea, Briſſ.
The Java Groſbeak, Lath.

The ſixth ſpecies is the Chineſe bird, deſcribed and figured by Edwards, and which he names *Padda*, or *Rice-Bird*, becauſe the Chineſe call rice in the huſk *padda*, which is the food of this bird. This author has painted two of theſe birds, and ſuppoſes, with great probability, that *Pl.* 41 repreſents the male, and *Pl.* 42 the female. We had a male of this ſpecies, which is delineated Fig. 1. No. 152. *Pl. Enl.* It is an exceedingly beautiful bird; for beſides the luſtre of the colours, its plumage is ſo perfectly regular, that no feather projects beyond another, but they appear covered entirely with down, or rather with a ſort of meal, ſuch as we perceive in plums, which produces a fine gloſs. Edwards adds little to the deſcription of this bird, though he ſaw it alive. He ſays only that it is very deſtructive among the plantations of rice; that the traders to the Eaſt-Indies call it *the Javan*, or *Indian Sparrow;* that this appellation would imply that it is found in the Eaſt-Indies, as well as in China; but he is rather diſpoſed to think that the

the Europeans, in their intercourſe between China and Java, had often carried theſe birds to that iſland; and laſtly, that what proves them to be natives of China is, they are painted on the Chineſe paper and muſlins.

The ſpecies which we are now to deſcribe are ſmaller than the preceding, and conſequently differ ſo much from our Groſbeaks, that we could hardly refer them to the ſame genus, did not the ſhape of their bill, the figure of their body, and even the order and poſition of its colours, indicate that theſe birds, though not exactly Groſbeaks, are ſtill nearer related to them than to any other genus.

XI.

The TOUCNAM COURVI.

Loxia Philippina, Linn. and Gmel.
Coccothrauſtes Philippenſis, Briſſ.
The Philippine Groſbeak, Lath.

The firſt of theſe ſmall foreign Groſbeaks is the *Toucnam Courvi* of the Philippines, of which Briſſon has given a deſcription, with a figure of the male, under the name of *the Philippine Groſbeak*, and which is delineated Fig. 2. No. 135. *Pl. Enl.* by that denomination. But we have here

here preſerved the name which it receives in its native climate, becauſe it differs from all the reſt. The female is of the ſame ſize with the male, but its colours are different, its head being brown, and alſo the upper-part of its neck, which in the male is yellow, &c. Briſſon gives alſo a figure and deſcription of their neſt *. [A]

* Theſe birds conſtruct a neſt of a ſingular form. It is compoſed of little interwoven fibres of leaves, which form a ſort of ſmall pouch, whoſe mouth is placed in one of the ſides. To this mouth is fitted a long paſſage, compoſed of the ſame leafy fibres, turned downwards, with its aperture ſituated below, ſo that the real entrance to the neſt is entirely concealed. Theſe neſts are faſtened by their upper part to the ſmall branches of trees. BRISSON.

[A] Specific character of the *Loxia Philippina*:—" It is duſky, " below yellowiſh-white, the top and breaſt yellow, the throat " duſky."

XII.

The ORCHEF.

Loxia Bengalenſis, Linn. and Gmel.
Paſſer Bengalenſis, Briſſ.
The Bengal Sparrow, Alb.
The Yellow-headed Indian Sparrow, Edw.
The Bengal Groſbeak, Lath.

The ſecond of theſe little foreign Groſbeaks is the Eaſt-India bird delineated *Pl. Enl.* No. 393. Fig. 2. under the name of *Indian Groſbeak*. I have termed it *gold-head (Orchef)* becauſe caufe the upper-part of its head is of a fine yellow, and being different from all the reft, required a diftinct name.—This fpecies is new, and has not been noticed by any preceding naturalift *.

* Specific character of the *Loxia Bengalenfis*:—" It is gray, " with a yellow cap; its temples whitifh, its lower-belly whitifh, " fpotted with dufky." It is fomewhat larger than the Houfe Sparrow.

XIII.

The NUN GROSBEAK.

Loxia Collaria, Linn. and Gmel.
Le Gros Bec Nonette, Buff.

The third of thefe little fpecies is that of Fig. 3. No. 393. *Pl. Enl.* which we have called *the Nun*, becaufe it has a fort of black biggen on its head. It is a new fpecies alfo; but we can fay nothing more of it, being unacquainted with its native climate. We bought it from a dealer in birds, who could give us no information on that fubject †.

† Specific character of the *Loxia Collaria*:—" It is yellowifh, " its breaft and collar yellow, its temples black."

XIV.

The GRAY GROSBEAK.

Loxia Grisea, Gmel.

The fourth of these is new, and as little known as the preceding. It is Fig. 1. No. 393. *Pl. Enl.* called *the Virginia Grosbeak*. But we shall term it *grey-white (Grisalbin)*, because its neck and part of its head is white, and all the rest of the body gray; and as it differs from the others, it merits an appropriated name*.

* Specific character of the *Loxia Grisea*:—" It is of a cœru-" lean gray, its neck and front white." It is four inches long.

XV.

The QUADRICOLOR.

The fifth of these little foreign Grosbeaks is the bird described by Albin, under the name of *the Chinese Sparrow*, and afterwards by Brisson †, under that of *the Java Grosbeak*, and delineated Fig. 2. No. 101. *Pl. Enl.* by the same name. We shall, however, term it *the Quadricolor*,

† The female, says this author, differs from the male, its thighs being of a light-chesnut, and the colour of its tail not so vivid.

dricolor, to diſtinguiſh it from all the reſt, and mark its principal colours; for it is a beautiful bird, and painted with four brilliant colours; the head and neck being blue, the back, the wings, and the end of the tail, green; there is a broad red bar, like a girth, under the belly, and on the middle of the tail; and laſtly, the reſt of the breaſt and belly is light-brown or hazel. We are ignorant of its habits.

XVI.

The JACOBINE, and the DOMINO *.

Loxia Malacca, Linn. and Gmel.
Coccothrauſtes Moluccenſis, Briſſ.
The Molucca Groſbeak, Lath.

The ſixth of theſe is the bird known to the curious by the name of *Jacobine*, which we retain as applicable and diſcriminating. It is repreſented *Pl. Enl.* Fig. 3. No. 139. and titled "The Java Groſbeak, called the Jacobine." We conceive that Fig. 1. of that plate, termed *the Molucca Croſsbill*, is of the ſame ſpecies, and probably a female of the firſt. We have ſeen theſe birds alive, and fed them like Canaries. Edwards deſcribes and figures them by the name

* Specific character:—"It is duſkiſh; its head, its throat, its "tail-quills, black; below it is waved with black and white."

of

of *Coury* *, Pl. XL. and from the meaning of this word, he infers that they inhabit India, and not China †. We would have adopted this term, had not that of *Jacobine* already come into uſe. Fig. 2. No. 139. and Fig. 1. No. 153. are two birds which the virtuoſi call *Dominos*, and which they diſtinguiſh from the Jacobines. They are ſmaller indeed, but ought to be regarded as varieties of the ſame ſpecies. The males are probably thoſe which have the belly ſpotted, and the females thoſe which have it of an uniform white-gray. The deſcription of them occurs in Briſſon's work, but not a word is ſaid of their natural habits.

* It is called a *Cowrie,* becauſe its ordinary price is one of the ſmall ſhells which paſs in India for money; but theſe are not current in China.

† It is the *Loxia Punctularia* of Linnæus, the *Coccothrauſtes Javenſis Nævia* of Briſſon. Specific character: "It is bay-coloured; its lower-belly black, ſpotted with white." It is four inches and a half long.

XVII.

The BAGLAFECHT.

Loxia Philippina, var. Gmel.

This is an Abyſſinian bird, much reſembling the *Toucnam Courvi;* the only difference conſiſting in the ſhades or arrangement of the colours.

lours. The black ſpot which is on both ſides of the head riſes in the Baglafecht above the eyes; the brown and yellow marbling of the upper-part of the body is leſs marked, and the great coverts of the wings and their quills; thoſe of the tail are greeniſh-brown, edged with yellow. Its iris is yellowiſh, and its wings, when cloſed. reach near the middle of the tail

The Baglafecht reſembles the Toucnam Courvi alſo in the precautions which it takes to ſecure its eggs againſt rain, and every ſort of danger; but the form of its neſt is different. The bird rolls it into a ſpiral nearly like the *Nautilus*, and ſuſpends it, as does the Toucnam Courvi, at the extremity of a ſmall branch, almoſt always above ſtagnant water, the aperture conſtantly turned to the eaſt, the quarter oppoſite to the rain. In this way the Baglafecht is not only ſheltered from the wet, but ſecured from the intruſions of different ſorts of animals, which ſeek to feed upon its eggs.

XVIII.

The ABYSSINIAN GROSBEAK.

Loxia Abyſſinia, Gmel.

I range among the Groſbeaks alſo the Abyſſinian bird, which reſembles them in the characteriſtic

racteriſtic feature, the thickneſs of its bill, and likewiſe in the ſize of its body. Its iris is red, its bill, the top and ſides of its head, its throat, and its breaſt, are black; the reſt of the under-part of the body, the thighs, and the upper-part of the body, light yellow, but which aſſume a brown tinge where the black of the anterior part meets it, as if the two colours there melted into one; the ſcapular feathers are blackiſh, the coverts of the wings brown, edged with gray; the quills of the wings and of the tail are brown, edged with yellow, and the legs reddiſh-gray.

The moſt ſingular fact of the hiſtory of the Abyſſinian Groſbeak, is the conſtruction of its neſt, and the ſort of foreſight which it diſcovers, in common with the Toucnam Courvi, and the Baglafecht. The ſhape of the neſt is nearly pyramidal, and the bird is always careful to ſuſpend it over the ſurface of water from the end of a ſmall branch; the entry is in the ſide, and commonly faces the eaſt; the cavity is divided by a partition into two compartments; the firſt is a kind of court into which the bird enters, then creeping along the incloſure, it deſcends into the ſecond chamber, where its eggs are laid. By means of this complex conſtruction, the eggs are ſheltered againſt the rain, from whatever quarter the wind blows: and we may obſerve, that in Abyſſinia the wet ſeaſon laſts ſix months: for it is a general remark, that inconvenience and hardſhip quicken induſtry, unleſs they be ſo exceſſive as to extinguiſh it entirely. In that country the bird was expoſed not only to the penetrating rains, but to the attacks of the monkeys, the ſquirrels, the ſerpents, &c. It ſeems to have foreſeen the dangers that threaten its family, and to have artfully provided againſt them. This ſpecies is new, and we owe all our information on the ſubject to Mr. Bruce.

XIX.

The GUIFSO BATITO*.

Loxia Tridactyla, Gmel.
The Three-toed Groſbeak, Lath.

There is no European ſpecies to which this foreign bird is more related than the Groſbeak. It ſhuns inhabited places, and lives retired in the unfrequented foreſts. It is languid in its amours, and deſtitute of ſong; and its only noiſe almoſt is made by the ſtrokes of its bill, in piercing the nuts to extract the kernel.—So far the analogy applies. But it differs from the Groſbeak by two remarkable properties; 1ſt, its bill is indented on the edges; and, 2dly, its feet have only three toes, two before and one

* The full name of this bird, as it is written in Mr. Bruce's drawings, is *Guifso Batito Dimmo Won Jerck*.

behind,

behind, which is an uncommon diſpoſition, and occurs only in a few ſpecies. Theſe two diſcriminating features ſeem to me ſo important, that the bird required an appropriated name, and I have preſerved that by which it is known in its natal region.

The head, the throat, and the fore-part of the neck, are of a fine red, which extends in a pretty narrow ſtripe under the body, as far as the lower coverts of the tail. All the reſt of the under-part of the body, the upper-part of the neck, the back, and the tail, are black; the upper-coverts of the wings brown, edged with white, the quills of the wings brown, with a greeniſh border, and the legs of a very dull red. The wings when cloſed reach not beyond the middle of the tail.

XX.

The SPOTTED GROSBEAK of the Cape of Good Hope.

The bird repreſented by this name, Fig. 1. No. 639. *Pl. Enl.* though different from the European Groſbeaks in its colours, and the diſtribution of its ſpots, appears ſo much a-kin to that ſpecies, that it may be regarded as a variety produced by climate, for which reaſon we have given it an appropriated name. And Sonnerat aſſures us poſitively that it is the ſame with that of the firſt article; and he adds, that theſe birds appear different, becauſe they change their colours every year.

XXI.

The CRAVATED GRIVELIN.

The bird delineated *Pl. Enl.* No. 659. Fig. 2. under the denomination of the Angola Groſbeak, becauſe we received it from that province of Africa, appears to be related to the *Grivelin*; and as all the neck and the under-part of the throat is covered and encircled by a ſort of white cravat, which extends even over the bill, we have given it the name of *the Cravated Grivelin*. We are ignorant of its habits.

The HOUSE SPARROW*.

Le Moineau, Buff.
Fringilla Domeſtica, Linn.
Paſſer Domeſticus, Geſner, Aldrov. Briſſ. &c.

As the ſpecies of Sparrow comprehends a multitude of individuals, ſo its genus ſeems at firſt inſpection to include a number of ſpecies. One of our nomenclators reckons it to contain no leſs than ſixty-ſeven different ſpecies, and nine varieties, making in all ſeventy-ſix birds †; among which we are ſurprized to find many Linnets, Finches, Green Birds, Canary Birds, Bengal Birds, Senegal Birds, Mayo Birds, Cardinals, Buntings, and many others not related to the Sparrows, and which ought to be diſtinguiſhed by ſeparate names. To introduce order into this confuſed group, we ſhall firſt remove from the Sparrow, with which we are well acquainted, all the birds juſt mentioned, which are alſo ſufficiently known to enable us to decide that they do not belong to the ſame genus. Following then our general plan, we ſhall conſider

* In Greek, Τρωγλιτης. Moſt tranſlators and naturaliſts have made it to be Στρυθος: the fact is, that this laſt name is generic, and applicable to all the Sparrows: in Italian, *Paſſera*, or *Paſſere Caſaringo*: in Spaniſh, *Pardal*: in German, *Huſſ-Spar*, *Haus-Sperling*: in Swediſh, *Taelting*, *Grawparf*.

† Briſſon.

each

Nº 85

FIG.1.THE SPARROW. FIG.2.THE MOUNTAIN SPARROW.

each of thoſe which inhabit Europe a principal ſpecies, and afterwards refer to them the analogous foreign kinds.

We ſhall alſo ſeparate from the Common or Houſe Sparrow the *Field Sparrow* and the *Wood Sparrow*; two birds more related than any of the preceding, and alſo inhabitants of our climate; to each we ſhall allot a diſtinct article. This is ſurely the only way to avoid confuſion.

Our Sparrow is too well known to need a deſcription. It is repreſented Nos. 6 and 55. *Pl. Enl.* fig. 1. No. 6. is the adult male after it has caſt its feathers; and fig. 1. No. 55. the young male before moulting. The change of colour in the plumage, and in the angles of the mandibles, is general and uniform; but the ſame ſpecies is ſubject to accidental varieties; for ſome Houſe Sparrows are white, others variegated with brown and white, and others almoſt entirely black *, and others yellow †. The only difference between the females and the males is,

* Black Sparrows occur in Lorraine; but they are undoubtedly the common ſort, which as they lodge conſtantly in the glaſs-houſes, which are frequent along the foot of the mountains, are ſmoked. Dr. Lottinger, being in one of theſe glaſs-houſes, obſerved a flock of common Sparrows, among which were ſeveral black ones. An old perſon, who dwelt on the ſpot, told him, that they ſometimes appeared ſo much diſguiſed, that they could not be diſtinguiſhed.

† Aldrovandus.

 that

that the former are ſmaller, and their colours much fainter.

Beſides theſe firſt varieties, ſome of which are general and others individual, and which occur in all the European climates, there are others in more diſtant regions; which would prove that this ſpecies is ſpread from the north to the ſouth in our continent, from Sweden * to Egypt †, Senegal, &c.

But in whatever country the Sparrow is ſettled, it never is found in deſert places, or at a diſtance from the reſidence of man. It likes neither woods nor vaſt plains. It is more frequent in towns than in villages; nor is it ſeen in the hamlets or farms that are buried in the depth of foreſts. It follows ſociety to live at their expence; and indolence and voracity lead it to ſubſiſt on the proviſions of others. Our granaries, our barns, our court-yards, our pigeon-houſes, and, in ſhort, all places where grain is ſpilt, are its favourite reſort. It is extremely deſtructive; its plumage is entirely uſeleſs, its fleſh indifferent food, its notes grating to the ear, and its familiarity and petulance diſguſting. In ſome places Sparrows are proſcribed ‡, and a price ſet on their heads.

* Linnæus. † Proſper Alpinus.

‡ In many villages of Germany, the peaſants are obliged annually to produce a certain number of Sparrows' heads. Frisch.

But what will render them eternally troublesome and vexatious, is not only their excessive multiplication, but their subtlety, their cunning, and their obstinacy to abide in places which suit them. They are crafty and artful, easily distinguish the snares laid for them, and wear out the patience of those who try to catch them. It is only in seasons of scarcity, and when the snow covers the ground, that the sport will succeed; and little impression can be made on a species which breeds thrice a-year. Their nest consists of hay, lined with feathers. If you destroy it, they will in twenty-four hours build another; if you plunder the eggs, which are five or six*, often more, they will in the course of eight or ten days lay others; if you drive them from the trees or the houses, they will resort in greater numbers to your granaries. Persons who have kept them in cages, assure me, that a single pair of Sparrows consume near twenty pounds of corn annually. We may judge from their numbers what prodigious destruction they must make in our fields; for though they feed their young with insects, and eat many themselves, they principally subsist on our best grain. They follow the labourer in seed-time, and the reaper in harvest. They attend the threshers at the barns, and the poulterer when he scatters grain to his fowls. They visit the pigeon-houses, and even

* Olina says, that they lay sometimes eight, and never fewer than four.

 pierce

pierce the craw of the young pigeons to extract the food. They eat bees, and are thus diſpoſed to deſtroy the only inſects uſeful to man. In ſhort, it is much to be wiſhed that ſome method could be deviſed for deſtroying them. I have been told, that if ſulphur were ſmoaked under the trees, where in certain ſeaſons they aſſemble and ſleep at night, they would be ſuffocated and drop dead. I have tried the experiment, without ſucceſs, though I took much pains, and was intereſted in the iſſue; for I could not get them driven from the neighbourhood of my voleries; and I perceived that they not only diſturbed the warbling of my birds, but that by the continual repetition of their harſh cry, *tui*, *tui*, they ſenſibly ſpoiled the ſong of the Canaries, Siſkins, Linnets, &c.

I then placed on a wall, covered with great Indian cheſnuts, in which the Sparrows aſſembled in great numbers in the evening, pots filled with ſulphur, mixed with a little charcoal and roſin; and theſe ſubſtances being ſet on fire, cauſed a thick ſmoke, which had no effect but to waken the birds. As the volume aſcended, they removed to the tops of the trees, and then retired to the neighbouring houſes, but not one dropped. I obſerved only that they did not for three days viſit the trees that were ſmoaked, but afterwards returned to their former habit.

As theſe birds are hardy, they can be eaſily raiſed in cages, and live ſeveral years, eſpecially if if the females be withheld from them *; for it is ſaid that their exceſſive venery abridges the period of their lives. When they are taken young, they are ſo docile as to obey the voice and catch ſomewhat of the ſong of thoſe birds with which they are bred; and being naturally familiar, they become more ſo in the ſtate of captivity. But when at liberty, they are rather ſolitary; and hence, perhaps, the origin of their name †. Since they never leave our climate, and are always about our houſes, it is eaſy to perceive that they commonly fly ſingle or in pairs. There are, however, two ſeaſons in the year when they aſſemble, not to fly in flocks, but to chirp together, in autumn on the willows by the river ſides, and in ſpring on the firs and other evergreens. They meet in the evening, and in mild weather. They ſpend the night on the trees, but in winter they are found either alone or with their females in a hole of the wall, or beneath the tiles of roofs. And it is only in exceſſive froſts that five or ſix are found lying together, probably to keep themſelves warm.

* "Some ſuppoſe that the male Sparrows cannot live longer "than a year. The proof is, that none are obſerved to have a "black beard in the ſpring, but only ſometime after, as if none "had ſurvived the preceding ſeaſon. It is alleged that the fe- "males are more vivacious; for they are caught along with young "ones, and are diſtinguiſhed by the hardneſs of their bills." Arist. *Hiſt. Anim.* lib. x. 7.

† Perhaps the French word *Moineau* is derived from the Greek Μονος, *ſolus*.

 The

The males fight obſtinately for the poſſeſſion of their females, and in the violence of their ſtruggle, they often fall to the ground. Few birds are ſo ardent, or ſo vigorous in their love. They can embrace twenty times in ſucceſſion with the ſame fire, the ſame trepidation, and the ſame expreſſions of rapture. What is ſingular, the female firſt ſhews a degree of impatience at a ſport which muſt fatigue her leſs than the male, but which may alſo yield her leſs pleaſure, ſince there are no preludes, no careſſes, no adjuſtment. Much petulance is ſhewn without tenderneſs, and a flutter of action which betrays only a ſelfiſh appetite. Compare the loves of the Pigeon with thoſe of the Sparrow, and you will perceive almoſt all the ſhades from the phyſical to the moral qualities.

Theſe birds neſtle commonly under the tiles, in the lead-gutters, in holes of the wall, in pots that are erected for them, and often about the ſides of windows which have Venetian blinds. A few, however, build their neſts in trees. I have received ſome of theſe which were found in large cheſnuts and lofty willows. They place them on the ſummit of theſe trees, and conſtruct them with the ſame materials, *viz.* hay on the outſide and feathers within ; but what is ſingular, they add a ſort of cap above which covers the neſt, ſo as to prevent the water from penetrating, and leave an opening for entering at under this cap. When they lodge in holes or covered places,

places, they judiciously dispense with this cap. Instinct discovers here a sort of reasoning, and at least implies a comparison of two small ideas. Some House Sparrows, more indolent, though bolder than the rest, do not give themselves the trouble of building, but drive off the Martins, and possess their nests. Sometimes they fight the Pigeons, and establish themselves in the holes.—This little tribe exhibit therefore habits and instincts more varied and perfect than most other birds. This results undoubtedly from their living in society. They enjoy the benefits of the domestic state without surrendering any portion of their independence. Hence that subtlety, that circumspection, and that accommodation of instinct to situations and circumstances. [A]

[A] Specific character of the Common or House Sparrow, *Fringilla Domestica*, LINN.—"The quills of its wings and tail "are brown, its body gray and black, with a single white stripe "on its wings." It is near six inches long. The eggs are ash-white, with thick dusky spots. The male is distinguished by his black throat; the female has a duller plumage. They occur throughout Europe, and in Africa and Asia.

FOREIGN BIRDS,

RELATED TO THE HOUSE SPARROW.

I.

The bird, delineated fig. 1. No. 223. *Pl. Enl.* under the name of *Senegal Sparrow*. We ſhall retain that denomination, ſince it appears to be of the ſame ſpecies with the Common Houſe Sparrow. The only difference is, that the bill, the top of the head, and the lower-parts of the body, are reddiſh ; whereas, in the European Sparrow, the bill is brown, the crown of the head, and the lower-parts of the body, gray. But in every other reſpect, they are the ſame ; and we may regard the difference of colour as reſulting from the influence of climate.

The bird of which the male and female are in fig. 1. and 2. No. 665. *Pl. Enl.* appears to be only a variety of this.

II.

We may extend theſe remarks to the bird fig. 2. No. 183. *Pl. Enl.* termed the *Red-billed Senegal*

Senegal Sparrow, which we ſhall conſider, eſpecially ſince it belongs to the ſame climate with the preceding, as a variety of it, occaſioned by difference of age or ſex.

III.

The BLACK SPARROW.

There are other foreign birds however, which, though analogous to the Houſe Sparrow, muſt be regarded as of a different ſpecies. Such is the American bird, which the inhabitants of the French Weſt-India iſlands call *the Black Father*, *(Pere noir)*. It is repreſented fig. 1. No. 201. *Pl. Enl.* It would appear to be ſettled not only in theſe iſlands, but on the continent of South America, as at Mexico; for it is mentioned by Fernandez under the Mexican name of *Yohualtototl*, and deſcribed by Sir Hans Sloane as a native of Jamaica *. We ſuppoſe alſo that the two birds, figured No. 224. are only varieties of this. The only thing which weakens this conjecture is, that they were found in climates very diſtant from each other: 1. from Macao, the 2d from Java, and the 3d from Cayenne. I ſtill conceive, however, that they are varieties of the Black Sparrow; for the climates allotted to them by the importers are not to be conſidered

* The Black Sparrow marked with ſaffron dots. SLOANE.

as

as certain; and besides, this species may occur equally in the hot countries in both continents.

There are others also which may be regarded as varieties of this species. The *Brazil Sparrow*, of which fig. 1. No. 291. *Pl. Enl.* is the male, and fig. 2. the female, resembles the Black Sparrow, so that we cannot hesitate to assign it the same place. The resemblance is indeed the most perfect in the male, for the female differs widely in its colours; but this circumstance only apprizes us of the uncertainty of any classification founded on the plumage.

Lastly, There is another species which we should range with the Black Sparrow, but for the great difference in the length of the tail. This bird is delineated fig. 1. No. 183. *Pl. Enl.* under the name of the *Sparrow of the kingdom of Juida*. We may consider it as a variety of the Black Sparrow, distinguished by its long tail, which consists of unequal quills. If we have been rightly informed with respect to the climates, it would appear that the Black Sparrow is found in the Antilles, in Jamaica, in Mexico, in Cayenne, in Brazil, in the kingdom of Juida, in Abyssinia, in Java, and as far as Macao; that is, in all the tropical countries, both of the New and of the Old Continent.

IV.

The DATE SPARROW*.

Le Dattier, ou Moineau de Datte, Buff.
Fringilla Capſa, Gmel.
The Capſa Finch, Lath.

Dr. Shaw ſpeaks of this bird in his Travels, under the name of *the Capſa Sparrow*, and Mr. Bruce has ſhewn me a miniature drawing of it, from which I have made the following deſcription:

The Date Sparrow has a ſhort bill, thick at the baſe, with ſome whiſkers near the angles of its junction; the upper-mandible is black, the lower yellowiſh, and alſo the legs; the nails black, the anterior part of the head and throat white, the reſt of the head, the neck, the upper, and even the lower ſurface of the body, gray, tinged with reddiſh; but the tint is deepeſt on the breaſt †, and on the ſmall upper-coverts of the wings; the quills of the wings and of the tail are black; the tail is ſlightly forked, pretty long,

* Mr. Bruce, after having attentively examined this bird, found it to be the ſame with the Maſcalouf of Abyſſinia. It is there called alſo *the Bird of the Croſs*, becauſe it uſually arrives the day of the Exaltation of the Holy Croſs, which denotes the cloſe of the rainy ſeaſon. Mr. Bruce adds, that at the ſources of the Nile a bird appears after the rains which reſembles much the Maſcalouf, except that it has a much ſhorter tail.

† Shaw ſpeaks of ſome reflexions which he perceived on its breaſt.

long, and ſtretches two-thirds beyond the extremity of the wings.

This bird flies in flocks; it is familiar, and ventures to pick up grains at barn-doors. In that part of Barbary, ſouth of the kingdom of Tunis, it is as common as the Houſe Sparrow in France; but it ſings much better, if what Shaw advances be a fact; that its warble is ſuperior to that of the Canaries and Nightingales *. It is a pity that it is too delicate to be carried out of its native country; at leaſt all the attempts that have hitherto been made of tranſporting it alive have proved unſucceſsful.

* I ſhould have been tempted from the fineneſs of its notes to range it with the Canaries; but Mr. Bruce, who had often ſeen it, and to whom I communicated my idea, perſiſted in his opinion, that it ought to be claſſed with Sparrows.

The TREE SPARROW*.

Le Friquet, Buff.
Fringilla Montana, Linn. and Gmel.
Paſſer Montanus, Aldrov. Ray, and Briſſ.
Paſſerinus, Geſner.
The Mountain Sparrow, Will.

THIS bird is undoubtedly of a different ſpecies from the Houſe Sparrow. Though they inhabit the ſame climate and the ſame tracts, they never aſſociate together, and their habits are, for the moſt part, diſſimilar. The Houſe Sparrow never leaves our dwellings, but lodges and breeds in the walls and roofs. The Tree Sparrow ſeldom viſits us, lives in the fields, haunts the ſides of the roads, perches on ſhrubs and low plants, and builds its neſt in crevices and holes at a little height from the ground. It is ſaid to neſtle alſo in the woods, and in the hollows of trees; but I have never ſeen them in the woods but tranſiently, and they certainly prefer the open fields. The Houſe Sparrow flies heavily, and always to ſhort diſtances; nor can it walk without hopping and making awkward movements. The Tree Sparrow, on the contrary, whirls round more ſmartly, and walks better. This ſpecies is not ſo numerous as that

* In German, *Baum Sperling*, *Feld Spatz*, or *Rohr Spatz*.

of the Houſe Sparrow; and it is exceedingly probable that they hatch only once a-year, laying four or five eggs; for about the end of ſummer they aſſemble in great bodies, and remain together during the winter. It is eaſy in that ſeaſon to catch them on the buſhes where they ſit.

After this bird has alighted, it is in a continual flutter, whirling, jerking its tail upwards and downwards, performing all theſe motions with tolerable grace; and hence comes its French name *friquet (friſky)*. Though not ſo bold as the Houſe Sparrow, it does not ſhun the preſence of man; it often follows travellers, without ſhewing any ſigns of timidity. It flies with a wheeling motion, and always very low; for it never perches on large trees, and thoſe who have called it the Cheſnut Sparrow, have confounded it with the Ring Sparrow, which really lodges on lofty trees and on cheſnuts.

This ſpecies is ſubject to variety. Many naturaliſts have reckoned the Mountain Sparrow *, the Collared Sparrow †, the Fooliſh Sparrow of the Italians, as ſpecifically different from it. But the Fooliſh Sparrow is exactly the ſame

* In German, *Ringel Spatz*, *Ringel Sperling*, *Feld Sperling*, *Wald Sperling*: in Poliſh, *Wrobel-leſmf*, *Wrobel-polny*, *Mazurek*.

† In German, *Berg Sperling*: in Poliſh, *Wrobel-garny*: in Greek, Σπεργιθος αγριος.

bird

bird, and the other ſorts are only ſlight varieties *.

What proves that the *Paſſera Mattugia* †, or Fooliſh Sparrow of the Italians, is either the Tree Sparrow, or a ſlight variety of it, diſtinguiſhed only by the diſtribution of its colours, is, that Olina, who gives a figure and a deſcription of it, ſays, that it receives the epithet of *Mattugia*, becauſe it can never reſt a ſingle moment in one place ‡; the ſame circumſtance to which I attribute the origin of its French name. Would it not be very ſingular, that this bird, which is ſo common in France, ſhould not at all be found in Italy, as our nomenclators have ſtated? On the contrary, it would ſeem that there are more varieties of this ſpecies in Italy than in France. It inhabits therefore the temperate and warmer regions, and not the cold climates, for it is not found in Sweden. But I am ſurprized that Salerne ſhould ſay that this bird occurs not in Germany or England, ſince the naturaliſts of theſe countries have given

* The Mountain Sparrow and Collared Sparrow are the ſame bird, and differ from the Tree Sparrow only by the white or whitiſh collar on the top of the neck.

† *Fringilla Stulta*, Gmel. *Paſſer Stultus*, Briſſ. *The Fooliſh Sparrow*, Will. and Lath.

Specific character:—" It is gray-rufous, ſpotted with ſooty, " below yellowiſh, its eye-brows and two ſtripes on its wings " white, a yellow ſpot on its throat, its tail blackiſh, and at the " margin rufous."

‡ *Paſſer Sylveſtris*, Aldrov.

figures

figures and deſcriptions of it. Friſch even aſſerts, that the Tree Sparrow and the Canary bird can breed together, and that the experiment has been made in Germany.

The Tree Sparrow, though more reſtleſs than the Houſe Sparrow, is not ſo petulant, ſo familiar, or ſo voracious. It is more innocent, and not ſo deſtructive to the crops. It prefers fruits, wild ſeeds, particularly thoſe of the thiſtle, and alſo eats inſects. It avoids meeting the Houſe Sparrow, which is ſtronger and more miſchievous. It can be raiſed in a cage, and fed like a Goldfinch; it lives five or ſix years; its ſong is very poor, but quite different from the harſh cries of the Houſe Sparrow. Though more gentle than the Houſe Sparrow, it is remarked not to be ſo docile. This is owing to its living more out of the ſociety of man *.

* Specific character of the Tree or Mountain Sparrow, *Fringilla Montana*:—"The quills of its wings and tail are brown, its "body gray and black, with a double white ſtripe on its wings." Its egg is white blue, with ſpots of a dull purple crowded at the thick end. It is found in Yorkſhire.

FOREIGN BIRDS,

WHICH ARE RELATED TO THE TREE SPARROW.

THE bird called the *Wild Sparrow* (*Paſſereau Sauvage*) in Provence, appears to be merely a variety of the Tree Sparrow. Its ſong, ſays M. Guys, would ſeem never to end, and is quite different from that of the Houſe Sparrow. He adds, that this bird is very ſhy, and conceals its head among the ſtones, leaving its body uncovered, and then fancies itſelf to be ſafe. It ſubſiſts in the fields upon grain, and ſome years it is very rare in Provence.

But beſides this and other varieties of the ſame ſort that inhabit our climates, and which we have mentioned after our nomenclators by the names of *Mountain Sparrow*, *Collared Sparrow*, *Fooliſh Sparrow*, there are others found in foreign climates.

I.

The GREEN SPARROW.

Le Paſſe Vert, Buff.

It is delineated fig. 2. No. 201. *Pl. Enl.* under the name of *Red-headed Cayenne Sparrow*.

We

We ſhall term it the *Green Sparrow*, becauſe its body is greeniſh. But though in point of colour it differs as much as poſſible from our Tree Sparrow, it is nearer related to this than to any other European bird.

II.

The BLUE SPARROW.

Le Paſſe Bleu, Buff.

The ſame may be ſaid of *the Blue Cayenne Sparrow* of fig. 2. No. 203.; and as both theſe birds inhabit the ſame climate, we can hardly decide whether they are diſtinct ſpecies, or ought to be ranged in the ſame.

III.

The FOUDI.

This bird is called in Madagaſcar, *Foudi Lehemené*. Briſſon mentioned it firſt under the name of the *Madagaſcar Cardinal*. It is delineated fig. 2. No. 134. *Pl. Enl.* by the title of *Madagaſcar Sparrow*.

There are two birds, the *Cardinal of the Cape of Good Hope*, fig. 2. No. 6. and *the Sparrow*

row of the Cape of Good Hope, fig. 1. No. 134. which both appear to me to be varieties of the Tree Sparrow, the former being the male, and the latter the female ; for the only difference is, that the under-part of the body is black ; but in all other reſpects they are alike, and as we have reaſon to believe that they live in the ſame climate, we may conclude they belong to the ſame ſpecies.

IV.

The CRESTED TREE SPARROW.

Le Friquet Huppé, Buff.
Fringilla Criſtata, Gmel.
The Black-faced Finch, Lath.

It is like the Tree Sparrow in ſize and ſhape, though much different in point of colour. It is delineated fig. 1. and 2. No. 181. *Pl. Enl.* under the names of the *Cayenne and Carolina Sparrow*. Fig. 1. is probably the male, and fig. 2. the female of the ſame ſpecies. [A]

[A] Specific character of the *Fringilla Criſtata*:—"It is creſted with duſky red, the under-ſide of its body, and its rump, "ſcarlet."

V.

The BEAUTIFUL MARKED SPARROW.

Le Beau Marquet, Buff.

It is delineated fig. 1. No. 203. *Pl. Enl.* under the appellation of *Sparrow of the Coaſt of Africa.* It is certainly different from the Tree Sparrow, and all thoſe which we have mentioned, and therefore required an appropriated name. That which we have formed denotes that it is beautiful, and finely ſpotted under the belly.

The

No. 36

FIG. 1. THE RING SPARROW. FIG. 2 THE GRIVELIN.

The RING SPARROW*.

Le Soulcie, Buff.
Fringilla Petronia, Linn. and Gmel.
Passer Sylvestris, Briss.
Passer Torquatus, Aldrov. and Ray.

THIS bird has, as well as the Tree Sparrow, been often confounded with the House Sparrow, though it is of a different species. It is larger than either, its bill is stronger, and red rather than black, and it has no habit in common with the House Sparrow. It dwells in the woods, and hence the name that it has received from most of the naturalists †. It nestles in hollow trees, lays four or five eggs, and hatches only once a-year. As soon as the young are able to accompany the parents, that is about the end of July, they associate in flocks. The Ring Sparrows are therefore collected six weeks earlier than the Tree Sparrows, and form also more numerous bodies. They remain united till the season of love, when they separate with their females in pairs. Though these birds are invariably stationary in our climate, it is probable that they dread the severity of the arctic region, for Linnæus makes no mention of them in his enumeration of the natives of Sweden. They

* In Italian, *Passara Alpestre*.

† Passer Sylvestris.

are

are birds of paſſage in Germany *, and do not arrive in flocks, but only one by one †; and what ſeems to confirm our conjecture, they are often found dead in the hollows of trees, in hard winters. They ſubſiſt not only on grain and ſeeds of all ſorts, but alſo on flies and other inſects. They are fond of the ſociety of their equals, and when they diſcover abundance of food, they invite them to partake. As they are almoſt always in numerous bodies, they do vaſt injury to newly-ſowed fields. They can ſcarcely be driven away or deſtroyed, for they partake of the caution of the Houſe Sparrow. They avoid ſnares, lime-twigs, and traps, but they can be caught in great numbers with nooſes ‡.

* This bird was hitherto ſcarcely, if at all, known in Germany; but of late years it has become very common.—*Note communicated by* LOTTINGER.

† Friſch.

‡ Specific character of the *Fringilla Petronia*:—"It is gray, "its eye-brows white, the upper-part of its throat yellow." Its egg is duſky, with white dots.

FOREIGN BIRDS,

THAT ARE RELATED TO THE RING SPARROW.

I.

The LITTLE RING SPARROW.

Le Soulciet, Buff.
Fringilla Monticula, Gmel.
Paſſer Canadenſis, Briſſ.
The Mountain Finch, Lath.

THIS bird is ſo much like the Ring Sparrow, that we might conſider it as only a variety, if it were poſſible that it could migrate into the New Continent. It is delineated fig. 2. No. 223. under the name of the *Canada Sparrow*. It is ſmaller than the Ring Sparrow, as all the American animals are inferior to thoſe of the ſame ſpecies in the Old World *.

* Specific character :—" It is brown, white below, its top variegated with gray-bay; its temples, its neck, and two ſtripes on its wings, white."

II.

The PAROARE

Is another beautiful bird, a native of South America. Marcgrave calls it by its Brazilian

 name,

[illegible]

[illegible] (fig. 2. [illegible]), [illegible] name of the [illegible] the King [illegible] animals are inferior to [illegible] or the [illegible]

[illegible]

[illegible]

[illegible] A N A R I [illegible]

[illegible] beautiful [illegible] South America [illegible] Brazilian name.

14

N.° 87

THE PAROARE.

name, *tije guacu paroara*, from which we have taken the term *Paroare*. Briſſon has named it the *Dominican Cardinal*, becauſe its head is red, and its body black and white. In the female, the fore-part of the head is not red, but yellow-orange, ſprinkled with reddiſh points.

We ſhall alſo apply the name of *Creſted Paroare* to a bird of the ſame continent, which appears to be only a variety, diſtinguiſhed by a tuft or creſt on its head. This beautiful bird is figured No. 103. *Pl. Enl.* and there termed *the Creſted Dominican Cardinal of Louiſiana.*

III.

The CRESCENT.

Le Croiſſant, Buff.

This bird is delineated fig. 1. No. 230. *Pl. Enl.* and there named *the Sparrow of the Cape of Good Hope*, which had been given to it by Briſſon. We ſhall term it the *Creſcent*, becauſe in its ſpecies and climate it is different from the others. In the diſtribution of its colours it is analogous to the Ring Sparrow, and has a white creſcent which extends from the eye below the neck.—This character is alone ſufficient to diſtinguiſh it.

END OF THE THIRD VOLUME.

Printed in the United States
By Bookmasters